ROAD RESEARCH

urban traffic models :
possibilities for simplification

**A REPORT PREPARED
BY AN OECD ROAD RESEARCH GROUP**

AUGUST 1974

ORGANISATION FOR ECONOMIC CO-OPERATION AND DEVELOPMENT

The Organisation for Economic Co-operation and Development (OECD) was set up under a Convention signed in Paris on 14th December, 1960, which provides that the OECD shall promote policies designed :

- *to achieve the highest sustainable economic growth and employment and a rising standard of living in Member countries, while maintaining financial stability, and thus to contribute to the development of the world economy;*
- *to contribute to sound economic expansion in Member as well as non-member countries in the process of economic development;*
- *to contribute to the expansion of world trade on a multilateral, non-discriminatory basis in accordance with international obligations.*

The Members of OECD are Australia, Austria, Belgium, Canada, Denmark, Finland, France, the Federal Republic of Germany, Greece, Iceland, Ireland, Italy, Japan, Luxembourg, the Netherlands, New Zealand, Norway, Portugal, Spain, Sweden, Switzerland, Turkey, the United Kingdom and the United States.

* *
*

FOREWORD

The Road Research Programme has two main fields of activity:

- promotion of international co-operation in road construction, safety and traffic, the co-ordination of research facilities available in Member countries and the scientific interpretation of the results of joint experiments;

- International Road Research Documentation, a co-operative scheme for the systematic exchange of information and of current research programmes in Member countries.

The present programme is primarily concerned with defining the scientific and technological basis needed to assist governments of Member countries in decision-making on the most urgent road problems:

- planning, design and maintenance of the total road infrastructure, taking account of economic, social and technical developments and needs;

- formulation, planning and implementation of common overall strategies for road safety;

- improvement of present traffic control systems (both on motorways and in cities) and the integration of existing and new transport systems and facilities.

o

o o

The Road Research Group on Simplified Urban Traffic Models was established in June 1972 to review the present state of development of urban transport models with particular reference to possibilities of simplification. The present report was finalised following four meetings of the Group in 1972 and 1973 which were held at OECD, Paris, the Technical University of Denmark, Copenhagen and the United Kingdom Transport and Road Research Laboratory. The report outlines the major features of simplified models and highlights the purpose of strategic models. Existing models are reviewed with special attention to the various components of strategic models and the implications for simplified models. The report indicates the major conclusions to be put forward to transport planners and presents recommendations for further work in the Group's subject area.

Investment decisions in the urban transportation sector reflect more and more changing emphasis from purely technical and economic solutions to broader system planning approaches involving the various sectors of the government and the community and taking into account human, social, economic and environmental factors and goals. In order to make urban transport investments and policies more effective, it is necessary to analyse and evaluate a wide range of alternative transport solutions before selecting those most adequate for detailed study. The aim of the Group was therefore to appraise and improve existing decision making tools available in the form of simplified traffic computer models.

The report focuses on the study of "strategic" models and provides information on the degree of simplification which is compatible with meaningful results. Many different mathematical models have been developed to represent the traffic conditions in urban areas. Such models normally consist of computer programmes which accept data describing a basic transport situation and hence may be used to estimate the effects on traffic of possible future changes. These models are simplified to the extent that they facilitate rapid analysis of many alternative transport policies, but to serve this purpose it is also necessary in some instances to retain considerable detail in particular aspects of the conditions studied, such as the use of a particular mode of transport or the group of the population which benefits or pays for a particular change.

It should be recognised that models suitable for strategic studies have particular characteristics. The basic requirement is that large changes in transport systems can be dealt with and for this reason it is usually necessary to study the equilibrium between supply of transport (in terms of infrastructure and level of service) and the corresponding demand. It is also necessary to be able to study changes in land use and the distribution of various activities. Examples of models with equilibrium between transport supply and demand have been developed but are comparatively rare, and need further development before they can be regarded as tools ready for general use.

It is important that the correct balance is struck between expenditure on the collection of data and on modelling and analysis. This is assisted by the fact that equilibrium models provide specific information on the accuracy of their results.

The transferability of data from one city to another deserves further consideration and improved international accessibility of transportation study data through a form of data bank may be beneficial.

It should be emphasised that strategic transport planning models are in their infancy and it is clear that an important step towards making them more useful for policy decisions would be to have improved criteria to define the quality of existing and proposed transport systems and services. It would be desirable to include environmental factors in such criteria. Further research is needed to refine present concepts to provide a widely applicable methodology and new ideas may well emerge in the near future. If world research on this fast-developing field can be brought together there is a promise that the urban policy-maker will be provided with vastly better information to assist his choice of planning objectives and investment decisions.

TABLE OF CONTENTS

CHAPTER V

DATA REQUIREMENTS, AVAILABILITY
AND SIMPLIFICATION

CHAPTER VI

SUMMARY, CONCLUSIONS AND RECOMMENDATIONS

APPENDICES

I

INTRODUCTION

I.1 Terms of Reference

The terms of reference for the Group were stated as follows:

Urban traffic remains one of the crucial problems of
governmental transportation policy. In order to make the
strategic choices in urban areas easier for those confronted
with present day problems (both in traffic management and
urban planning), a study of the applicability of simplified
techniques in the formation of transport planning in urban areas
will be attempted with special emphasis on road networks.
In this context, it will also be necessary for the Group
to consider such problems as traffic generation, distribution
assignment and modal split.

The objectives will be:

- to review the present state of development of urban transport models with
particular reference to:

i)   their complexity
ii) their validity

- to consider the minimum data required for an acceptable model both for present
and future requirements;

- to report on the existing urban transport models to real situations;

- to specify future research needs.

I.2 Reasons for Simplification

These terms of reference reflect the fact that many transport planning authorities
need means of appraising possible investments or policies without detailed transportation
studies of the classical type which involve the collection and analysis of large volumes
of data and the running of complex and costly computer models. This process is unwieldy,
inflexible and slow and it is not certain that the level of detail and accuracy is well
matched to the planning phase concerned.

This situation has given rise to a number of "strategic" transport models which
enable the user to study a wide range of possible alternatives before picking the most
likely ones for detailed study. Less affluent towns may limit their planning to the
strategic level and it may be possible to prepare a sufficient background of planning
guidelines to minimise individual detailed studies in order to reach the required conclusions in some instances.

I.3 Field of Interest

A definition of the class of transport planning models considered by the Group
has evolved as follows:

i) The model should be suitable for strategic studies in that broad policy effects should be studied rather than detailed projects. Since results must be intelligible, the models must be clear and understandable and must be suitable for instructional use.

ii) It should be possible to apply the models without expensive special-purpose (home interview) surveys. Particular attention should be paid to refurbishing or generalising existing data and the study of home-work trips only may suffice in some cases.

iii) The period to be studied would range from the near future to, say, 15 years or more hence; models of a sector or subarea of the overall transport system often deal with the shorter term predictions.

iv) The relationships between model cost (and hence complexity) and the necessary and obtainable accuracy should be understood (developments in methodology have an important bearing on this).

v) Clear and fully analysed results on a planning alternative should be available within a short time to permit the study of a wide variety of alternatives. The importance of effective and imaginative presentation of results should be emphasized.

The above points have been made on the overall understanding that the cost of this phase of modelling should be minimised and that there is a need for parallel activities on both strategic and detailed models, certainly for the larger urban areas, the strategic results being normally available first.

In the sector model mentioned in the third item above, interest may be directed towards a particular corridor of movement, a particular mode of travel, etc., narrowing the coverage of the model but perhaps incorporating considerable detail in the sphere of interest, for example with a view to studying the effect of adding a single new line of communication or new activity centre to an existing situation. This report does not deal with this type of model but it may be an appropriate subject for an international review.

Discussion in the Group has indicated that the key question regarding the strategic type of model lies in the degree of simplification which is compatible with meaningful results; this is a delicate compromise on which more information is needed. The more familiar tactical type of model which is used to study the traffic engineering questions of a given transport network is not free from accuracy problems but they are less dominating than in strategic models.

I.4 Current Situation

Members of the Group were asked at the outset to contribute statements of the status of national work on this general class of model. This remit intentionally permitted a variety of interpretations of the model type with a view to establishing a consensus of both interest and current practice before focusing on what emerged as the main problem areas. The outcome of this process is summarised in Chapter III.1.

As an example of typical planning practice, the three phases of planning studies used in France define a phase I which describes the aims of long-term strategic planning, i.e. the aim is to assess transport needs as they stem from long-term town planning objectives under different assumptions, design, and implied transport infrastructure and hence to indicate a coherent long-term (1990-2010) town planning and transport infrastructure scheme. This is followed by a second phase which designs the resulting projects in sufficient detail to reserve rights of way, etc. A special study is made of traffic flow at key points, the effects of network changes and the costs involved. The third phase is the firm five-year programme of infrastructure additions adopted.

The broad outline of this process is similar to that adopted in other countries (1,2) but the explicit identification of phase I as appropriate to the application of a particular class of simplified traffic model closely matches the field of interest of the Group.

I.5   The Function of Models in Urban Planning

It is clear that the overall process must include examination of transport, environment and land use effects. Each of these requires models which permit a detailed examination of future situations to enable planners to choose the best overall compromise. The process involved is illustrated in Figure 1.

Figure 1

ROLE OF MODELS IN PLANNING PROCESS

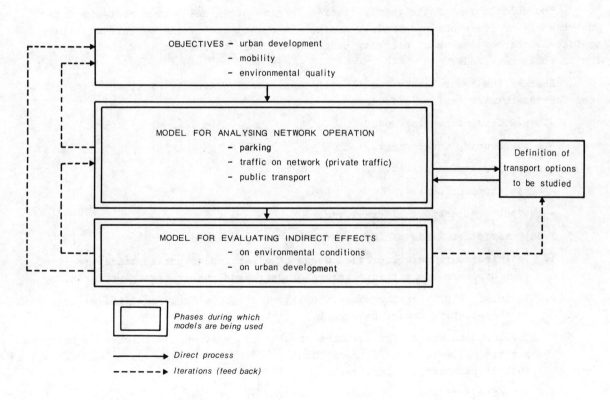

9

Environmental effects and social repercussions may perhaps be considered outside the sphere of traffic models. It is true that limited progress has been achieved in quantified modelling of such aspects but the importance of these questions demands that urban transport modelling must be adapted to be compatible with their evaluation and any long-term studies should have provision for outputs and inputs at the interface between traffic and social effects. There is, however, a considerable lack of understanding of the interaction between transport facilities and land use patterns. Practice at present relies upon a fixed land-use plan used as the input to a traffic study. An alternative land use plan may then be introduced for a repeat study. Clearly it is possible to examine only very few alternatives in this way. These questions are discussed in section III.2. The subject of environmental effects of traffic has been dealt with by another O.E.C.D. Group (3) and by some projects of the O.E.C.D. Environment Committee, but the relevance of the traffic models considered below is an important aspect of their usefulness.

In practical terms, models have generally been used to calculate future traffic on the road network. This calculation shows the bottlenecks and gives an idea of where the road network should be improved.

In some cases several land-use and traffic plans are prepared at once, and traffic is calculated for all combinations of land-use and traffic plans. In other cases the first anticipated land-use plan gives rise to special traffic difficulties and a new land-use plan and traffic plan is prepared.

The calculation of the yearly traffic is also often used for a cost-benefit analysis of the different proposals. Another important use of the calculated future traffic is the setting up of priority programmes: when to build the various phases of the proposed new roads.

Besides the comprehensive transport plans covering a whole city, the models are used for determining traffic effects of:

- a new circumferential road,

- a new housing development,

- a new pedestrian street,

- a new shopping centre,

- new public transport alternatives.

Today many other kinds of issues are also being studied:

1. Policies to promote increased usage of a particular public transport line, e.g. by increased parking charges or decreased public transport fares.

2. Policies to provide increased mobility for special groups of travellers, such as the elderly, young, handicapped, etc.

3. The location and design features of alternatives, such as highways, improved arterial streets, traffic management schemes, and parking policies, as well as public transport alternatives.

4. The effect of staggered working hours on street traffic.

5. The effect of traffic policies intending to reduce energy consumption, air pollution and noise.

6. The development of new public transport alternatives, e.g. dial-a-bus.

In connection with the broad view of the position of transport planning in the overall urban planning process, such questions as the relative benefits derived by different social groups, the desirability of, say, public transport systems which may not top the list of cost-benefit returns on the conventional user benefit basis, etc., raises the question of the basis of evaluation. The four main themes of evaluation criteria relate to transport utility, system costs, environmental (and aesthetic) costs and land use effects. These cannot at present be related by a common set of units but the importance of each is not to be overlooked because only one aspect is dealt with fully in this report.

I.6   Features of Simplified Models

Important features of such models should be the ability to communicate to those responsible for policy and administration such ideas as the link between future land use, residential density, employment, etc., and the need for commensurate transport infrastructure(4,5). This is of critical importance in the case of the city centre, and has to be viewed in the context of the potential growth in traffic.

The models should be easy to operate and therefore input data requirements must be limited and suitable for reliable extrapolation (e.g. population and employment densities). Too much refinement, such as coefficients resulting from multiple regressions, can be self defeating if they do not represent understandable phenomena(6).

The question of travel mode is fast becoming the key to urban transport problems. Most cities can envisage an adequate transport network if public transport service levels and patronage can be improved. In most cases this can be done with currently available hardware given improved infrastructure, particularly interchange facilities, and possibly some private transport restraint. However, the maintenance of off-peak services and high density peak requirements with increasing labour costs make more use of automatic systems attractive. Models are needed which will assist in the evaluation of these prospects.

The model types which have so far resulted from these common broad objectives are described in this report. These represent different choices in the degree of simplification which is adopted in the model components. Decisions are also necessary on time effects, both with regard to peak/off-peak conditions and as to whether the history of changes with time should be reproduced, the latter implying more complex modelling than would be appropriate for strategic studies by present methods.

Examples of simplified models described in the report represent various choices in the representation adopted in the components listed below:

1) The number and type of zones.
2) The population characteristics.
3) Trip generation factors, effects of land use and journey purpose.
4) The transport network, the balance between public and private transport facilities.
5) The modes of transport and factors affecting choice of mode.
6) The objective functions or definitions of utility and the evaluation process.
7) Travel decision criteria (including route and destination alternatives).
8) Environmental effects.
9) Social, commercial and political repercussions, impact on land use.

The work of the Group aims to clarify the relationship between the choice of simplification and the usefulness of the resulting models, to assess the demand for models of particular types and to pinpoint the features on which research into methodology or input data are especially worthwhile. It will be noted that the above list does not clearly separate the conventional stages of trip generation, distribution, mode split and assignment (conventional "four-step" models) since in some cases improved modelling has been achieved by combining these processes in an "equilibrium" model (see section II.4) The applicability of such equilibrium models compared with four-step models will be examined.

An attempt has been made to assemble the types of model discussed in this report into a systematic summary. It is difficult to do this without an overwhelming amount of detail but table 1 describes the leading particulars of interest. Figure 2 shows a plot of two of the variables in this table and is useful in indicating two main themes; tactical (few alternatives, more detail) and strategic (variety of alternatives, less detail) models. This diagram shows a middle ground where the divergence of these two classes is negligible but it can be said that geographical simplicity is a common theme in models designed for strategic studies; it is usual for details of modes and users to be retained for such purposes. Section V.2 outlines some ways for simplifying tactical models.

It has been found that these strategic types of model have been necessary for the study of policies for short term application (e.g.: measures to meet fuel shortage). To summarise the usual applications of the main model types:

Type	Section	Application
Four-step model	II.3	Traffic and transport studies in well defined situations with extensive data coverage.
Simplified four-step model	App.3	Studies with limited data requirements for circumstances which do not vary greatly from a base condition.
Equilibrium demand models	II.4	Well suited to strategic studies particularly in cases of important changes or of traffic restraint.

Table 1

Summary of Urban Traffic Models

TYPE	PURPOSE	NETWORK	PERIOD (YRS)	AREA	INPUT DATA	BASIC STRUCTURE	USER CLASSIFICATION	NON-USER EFFECTS	OUTPUTS	CUSTOMER
(1) Detailed 4-step e.g. SELNEC HRB TAP FABER	Overall transport planning	Geographic all main routes	10-20	Conurbations and large towns	Home interviews Traffic surveys PT surveys Land use Growth factors Generation assumptions	Generation Distribution Mode split Assignment (capacity restraint) Evaluation	Car availability Trip purpose Time value	Model outputs suitable for environmental assessments	Traffic link Speeds and flows Mode patronage User benefits	Local Authorities
(2) Coarse 4-step e.g. COMPACT IMPACT FABER	Evaluation of major planning Alternatives (strategic)	Geographic-skeletal	10-20	" Also smaller towns	Simplified trip matrix Traffic counts Generation assumption	As above but small no. of zones	Permits Car/P.T. split but user largely aggregated	Results could provide background environment studies in chosen zone	Corridor speeds and flows Broad P.T. patronage Major effects on user benefits	Local Authorities Policy studies
(3) Hierarchical e.g. SALTSFORD LINCOLN LTS ZONE 277	Detailed study of chosen small area with effect of surrounding network	Geographic-detailed surrounded by skeletal	5-15	Smaller towns and chosen zones of conurbations or large towns	Local survey data in zone surrounding data from (2) or CORSS etc.	As for (1) but cordon traffic + internal trips provide basic pattern	Suitable for study of trip purpose and user class effects	Provide data for environment study in chosen zone	As for (1) in chosen zone Could input to e.g. junction design	Road authorities and policy studies by example
(4) Idealised e.g. CRISTAL RELTAP (circular)	Strategic and policy studies	Geometric skeletal e.g. ring/radial	5-20	Large towns	If data for (1) available realism improved but trends can be studied using synthetic data	Demand/supply equilibrium models Considerable detail of modes	Mainly by elasticity time valuation and car availability	Could be used for input to (3)	Corridor speeds and flows. Traffic generation and mode split, user benefits and P.T. effects	Policy studies with scope for particular examples
(5) Sector e.g. WMPT Systems Assess COBA	Study of alternative projects to meet a need shown by (1)-(4)	Minimal in some cases a single corridor	0-15	Any	Output from (1)-(4) with access to effects of alternative projects	Mainly mode and route choice formulation with cost and benefit evaluation	Level of aggregation depends on objectives	Not well suited to overall study but local effects could be derived	Performance of chosen routes but network effects largely lacking	Public transport authorities Road building authorities
(6) Single Link	Spot checks of effects of particular parameters	Homogeneous	0-10	Any area which can be regarded as homogeneous - usually large urban areas	General average values	Single link characteristics only Network and distribution effects not included but a form of capacity restraint possible	User characteristics can be included as an assumption or a parameter	Not appropriate	Useful quick check on effect of particular parameters but effect of input constraints on evaluation need care	Preliminary studies in connection with types 1-5
(7) Land-use interaction e.g.	To relate transport needs to planned land-use	Simplified but including all public transport	10-30	Particularly developing towns and areas with changing employment distribution	Existing and planned network Employment locations and residence by class of worker	Type (2), (3) or (4)	User disaggregation usually essential Work trips first priority but residential etc. becoming important	Should be a major component of evaluation	Accessibility is currently widely used but criteria describing the special function of transport should be developed	Urban Planning authorities

Figure 2

MODELS CLASSIFIED ACCORDING TO GEOGRAPHICAL DETAIL AND TIME HORIZON *

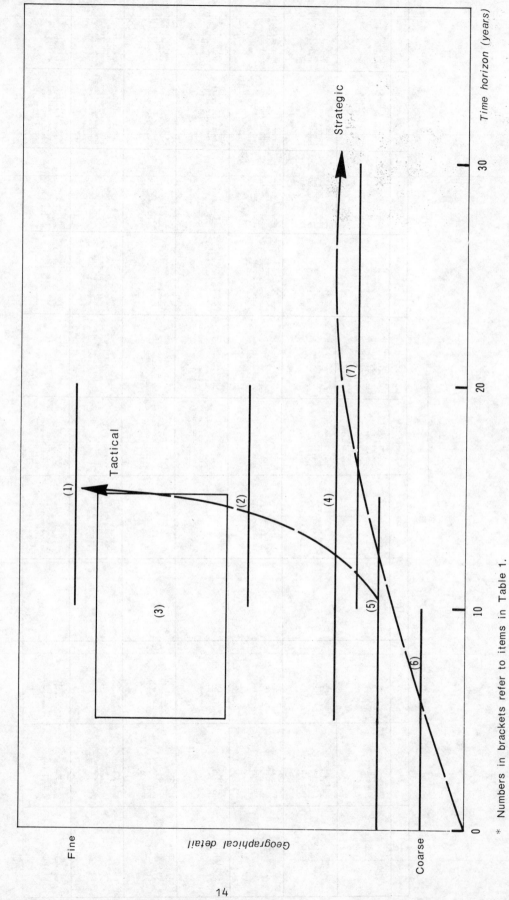

* Numbers in brackets refer to items in Table 1.

14

II

EXISTING MODELS

II.1 Introduction

This chapter indicates the current state of the art in urban traffic models, i.e. the four-step models conventionally used in urban transportation and the new single-step "explicit" demand models. This review is necessary in order to provide a background for the consideration of possible simplification methods.

The basic theory which underlies travel forecasting techniques will first be reviewed, then, the concept of the General Share Model will be introduced, and used to show the relationships between the conventional models and the equilibrium models. The relationship of disaggregate and aggregate modelling approaches will be described. Mention will be made of urban dynamic development models. Finally, the implications for developing simplified traffic models will be discussed.

II.2 Theory

II.2.1 Statement of problem

The problem of predicting the flows in a transportation system is an application of economic theory: the flows which will result from a particular transportation system (T) and pattern of socio-economic activities (A)* can be determined by finding the resulting equilibrium in the transportation market. If:

V = volume of flow

L = level of service experienced by that volume

F = (V,L) = flow pattern

then, an equilibrium is found by establishing a supply function (S) and a demand function (D), and solving for the equilibrium flows (F_0) consistent with both relations(7,8,32):

$$\left\{\begin{matrix} L = S(V,T) \\ V = D(L,A) \end{matrix}\right\} \longrightarrow \left\{ F_0 = (V_0, L_0) \right\}$$

This is illustrated in figure 3, Basic Theory; with travel time, t, as the level of service, L.

---

* The pattern of socio-economic activities (A) denotes the magnitude, spatial distribution, and composition of population, employment, land use, social structures, economic activities and any other aspects of human society which may influence the demand for transportation.

Figure 3

BASIC THEORY

(See text for symbols and formulae)

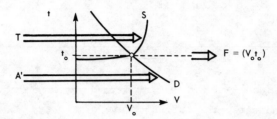

While simple in outline, the application of the theory becomes complex in practice for several reasons:

a) the consumer considers many service attributes of the transport system when making a choice (e.g., line-haul travel time, transfer time, walk distance, out-of-pocket cost, privacy, etc.), and thus, L must be a vector with many components.

b) determining the demand functions (as well as other elements) to use is difficult(32);

c) the equilibrium occurs in a network, where flows from many origins to many different destinations interact, competing for the capacity of the network; and the form of these interactions is affected by the topology of the network.

Thus, fairly elaborate computational schemes are required to actually determine the equilibrium flows F_O for a particular (T,A).

In the case of a multimodal network, the symbol V represents an array of volumes and the aim of the equilibrium procedure is to obtain the volumes and the level of services experienced for all journeys and modes.

Unfortunately, at this state of the science of transportation modelling, while several systems of transportation models exist, there is not even one operational model which solves for these equilibrium flows exactly and directly.

II.2.2 Alternative approaches

One particular computational scheme is that used in four-step urban transportation planning studies. In this indirect approach, the equilibrium flows are estimated in a sequence of steps, commonly called trip generation, trip distribution, modal split, and traffic assignment (e.g. 9,10).

Correspondingly, the demand function, D, is represented as a sequence of functions: trip generation equations, trip distribution procedures (including the "friction factor" transformation of L), modal split equations, and the minimum-path rules of the traffic assignment procedures.

More recently, alternative approaches have been developed. One approach uses explicit demand models to estimate the equilibrium flows in a direct approach, in a single step instead of the sequence of steps as in the indirect approach. Such explicit demand models combine the functions of generation, distribution, and modal split (and potentially, route choice) into a single process. The first such models were developed for forecasting intercity passenger travel(11-15). Later work extended these models to urban travel(16-19).

There are a number of alternative approaches which are or can be taken to predict flows in networks as the equilibrium of supply and demand. Each approach involves specific assumptions, both explicit and implicit, in the choice of demand models(20-22) and of the computational procedures for determining equilibrium. Very serious biases may occur in the computed flow patterns, as compared with the "true" equilibrium, if the assumptions and computational approaches are not carefully considered.

In developing simplified models for traffic forecasting, further assumptions and approximations are required. In order to understand the implications of such simplifications, it is important that the assumptions of present techniques be clearly understood.

II.2.3 Relationship between conventional and equilibrium models

By introducing a "general share model" it is possible to prove that the conventional four-step model is a special case of an implicit demand model. By doing this, it is also possible to place the two - direct and indirect - approaches in their proper context. As can be seen in Appendix 1, the general share model itself is an explicit demand model, in that the volume V appears only on one side of the equation

$$V = \psi(Y), \text{ where } \psi \text{ is a demand function.}$$

The notation Y denotes the set of variables which fully characterise the level of service offered by a transportation system, and the pattern of socio-economic activities. For example, in the simple gravity model Y denotes the populations in each zone (P_k), the employment in each zone (E_k), and the travel time matrix (t_{kl}).

In the more general case, of an implicit demand model, volume V appears on both sides of the equation:

$$V = \psi(Y, V)$$

Of particular importance is a special case of implicit demand model, the sequential implicit model which can be written as shown in Appendix 1.

Because the general share model is an explicit form of demand model, it can be used in a direct approach to computing equilibrium (i.e., a single computational step). Because it can also be expressed in an equivalent sequential implicit form, it also can be used in an indirect approach to computing equilibrium (i.e. a sequence of steps - generation, distribution, etc.).

II.2.4 Concluding remarks

These theoretical considerations place the two main classes of existing models within a systematic framework. The Group recognises the value of the general share model as outlined above since it constitutes a general basis for assessing the underlying assumptions of four-step models and for improving present modelling techniques used in urban transportation planning. Furthermore, it is a useful tool when developing simplified models. These implications are dealt with in section II.6 of this chapter.

II.3 Four-Step Models

II.3.1 Introduction

In the section on theory, the differences between four-step and demand models were briefly outlined. Both model types aim at the same result: the prediction of person or vehicle flows on parts or all of the transportation system. While demand models set out to achieve this result in a single set of mathematical calculations, four-step models employ the stages generation, distribution, modal split and assignment to arrive at the flows.

The models derive their general name from the presence of the above four distinctive steps. The steps form a logical sequence as illustrated in figure 4. It must be emphasised, however, that many different sequences and feedback loops are possible, depending on the level of complexity (or simplification) adopted.

Figure 4

SIMPLIFIED OUTLINE OF FOUR-STEP MODELS

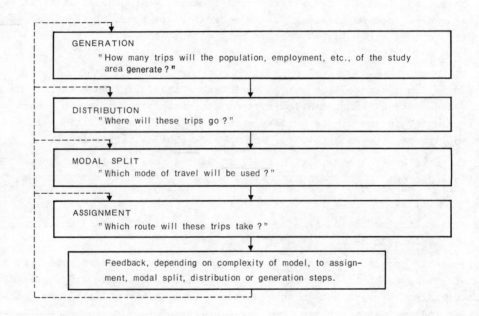

Being largely self-contained, each step lends itself well to computerised analysis. Moreover, within the limitations of existing computer techniques such a segmented analysis procedure can be easily extended to handle large urban areas in considerable detail.

The key question, and one which prompted the establishment of this Group, is whether this increased level of detail, complexity and cost is justified by a corresponding increase in the accuracy and validity of the models relative to what could be achieved by a more simplified approach.

The remainder of this section of the report concentrates on describing important aspects of existing four-step models in order to provide a background against which simplified models can be assessed.

II.3.2 Trip generation

The trip generation stage of a conventional four-step model normally consists of a series of mathematical formulae designed to answer the question "given a specified disposition of land uses, how many trips will start from or end in specified areas or zones in the study area?" In this context, a trip may be defined as a one way person (or vehicle) movement from a given origin (e.g. home) to a given destination (e.g. work place).

Hitherto, it has been usual to develop new trip generation formulae for each urban area studied. Much reliable data are required to establish the parameters of the formulae. Most of the data are generally collected from home interviews, that is, interviews conducted in a representative sample of households in the area. The remaining data - on travel not generated by households (for example, truck trips), on land uses and on transportation network conditions - are collected in supplementary surveys.

The development of a traffic forecasting model based on analysis of an extensive home interview survey has been the characteristic of classical four-step models. Basically, the "home interview analysis" is used for

- calculating the number of trips generated in each zone or household
- finding the "constants" for generation of traffic
- finding the present number of trips between zones
- determining the "travel law" or relationship between travel and land use appropriate to the study area.

Likely future changes in the land use plan may modify the travel pattern of the area and account must be taken of this in the travel forecasts.

It is usual to compare the travel found in the surveys and predicted by the traffic forecasting model with the actual traffic crossing several "screen lines" drawn across the city.

The normal method adopted for establishing the parameters necessary to give the "constants" for traffic generation is to carry out some type of regression analysis on the home interview and other survey data. This regression can be carried out in one of three ways - at a zonal level, at a household level or by cross classification.

Zonal regression analysis has been used widely - especially in early transportation studies. In it, each zone is treated as one unit of analysis. The average number of surveyed trips entering (or leaving) the zone for some particular purpose by a specified mode of travel is taken as one value of the dependent variable, and zonal average values for such items as income, car ownership, number of households, number of persons employed, etc., are taken as one corresponding set of values for the independent variables. The constants of a regression equation relating the dependent and independent variables are then calculated, using sets of survey data from all zones. With this technique, it is necessary to have sufficient survey observations in each zone to provide reliable estimates of the average zonal value of the dependent and independent variables.

Zonal regression suffers from the disadvantages that a linear relationship is usually assumed between the dependent and independent variables, that the independent variables are assumed to be uncorrelated with each other (and therefore additive) and that all variables are assumed to be continuous. Difficulties are also introduced by zonal aggregations which can mask vital effects or introduce bias according to the local conditions(39).

Household regression analysis is similar in operation to zonal regression, except that each household surveyed is treated as one unit of analysis, so that a smaller sample size suffices. Apart from the fact that it has been largely ignored as a trip estimating technique, it suffers from the same disadvantages as zonal regression.

Cross classification, though not, strictly speaking, "regression analysis" is, nevertheless, an application of some of the techniques of regression analysis. In it no assumption of a linear dependent/independent variable relationship is made. Cross classification consists of calculating from survey data, the average value of the dependent variable for each of a number of classes, or groups of values of the independent variables. For example, one class might consist of the average number of work trips generated by medium density high income one car owning households. It is thus apparent that a disadvantage of cross classification is the need to ensure that the surveys collect enough data to permit the calculation of meaningful average trip rates for each of the selected classes of households.

Every trip has both a starting point and an ending point, and it is thus normally necessary to develop separate sets of generation equations to predict the numbers of trips starting from, and ending in each zone. However, equations such as those outlined above are usually reserved for "internal" trips (trips with both ends within the study area). Less elaborate techniques are used for "external" trips (where one trip end is inside the study area and the other is outside) and "through" trips (both ends outside the study area).

It is necessary to recognise that seemingly precise mathematical formulae which have been developed for trip generation models may still embody a considerable amount of data specific to individual studies. (See section V.1.4.)

The presence of many independent variables in conventional trip generation models implies a need for large amounts of survey data for model development. This, in turn, implies costly and time consuming study processes. Much of the work described in sections III.1 and V.2 has set out to simplify conventional models through a reduction in the amount of data collected.

II.3.3 Trip distribution

Trip distribution models are used to estimate the number of trips made between pairs of zones once the total number of trips starting from and ending in each zone is known, and once some index has been established which expresses the efficiency of the transportation network connecting the zones. Two main groups of models have been developed: Growth Factor and Synthetic.

The basic philosophy underlying growth factor models is that present zone-to-zone movements can be projected into the future solely on the basis of anticipated rates of growth in the zones' total generations and attractions. In their simplest form, they use a single factor to scale existing zone-to-zone movements (found by survey) to future movements. More complex versions attempt to incorporate both the absolute and relative growth rates of different zones.

Synthetic models attempt to identify some of the factors which influence zone-to-zone movements(10,24). This is done by analysing existing movements (found by survey) and modifying the parameters of the model until it can predict the present movement pattern to an acceptable degree of accuracy - a procedure known as "calibration." These parameters are assumed to remain constant with time, and in this way the model can be used to predict future movements irrespective of the extent of changes in land use.

The general formula adopted for synthetic distribution models (ignoring purpose or mode of travel) is:

$$tij = p(i) \ a(j) \ f(i,j)$$

where

tij	=	trips produced in zone i and attracted to zone j
p(i)	=	factor depending on the production zone i
a(j)	=	factor depending on the attraction zone j
f(i,j)	=	factor representing the conductance (e.g. generalised cost) between zones i and j.

Appendix 2 gives examples of the Gravity Model formula (a special case of the above), together with a review of distribution model formulae as used in various studies.

II.3.4 Modal split

Modal split models attempt to indicate the percentage of trips which will be made for each purpose by the various modes considered(25). They fall into two main groups: Pre-distribution and Post-distribution.

Pre-distribution models involve determining the proportions of the productions and attractions in each zone which can be ascribed to each mode. Separate distribution and assignment models are then used for the modes.

Post-distribution models divide the zone-to-zone person trips, obtained from the generation and distribution models, into person trips by each mode. Separate assignment models are then used for the modes.

The main disadvantage of pre-distribution models is that they cannot easily include any measures of transport system efficiency in their formulae. However, post-distribution models in effect assume that the total number of zone-to-zone person trips is independent of mode, and that the proportion of these trips made by each mode can subsequently be determined from relative travel times or costs on the modes.

In the case of post-distribution models applied to two mode (car and public transport) problems, diversion curves are often constructed which relate proportion of trips made by public transport to the ratio of public transport to private transport travel times. Different curves may be used depending on the parking conditions in the generating and attracting zones, since parking has been found to have considerable influence on modal split.

Other factors which have been taken into account include:

- average income of inhabitants in the zone

- relative travel costs (including car parking charges)

- service ratio, i.e. the ratio between "time spent walking and waiting for public transport" and "time spent walking and parking for private transport."

II.3.5 Trip assignment

These models are used to calculate the number of persons using each link of the public or private transport network - figures required both for network evaluation and for design(26).

In its simplest form assignment involves determining the route of the shortest (or lowest cost) path between each pair of zones, adding to the volumes on every link of this route the number of trips given by the distribution model for this zone-to-zone pair, and repeating the process for all other zone-to-zone pairs. This is known as "all-or-nothing" assignment, since all trips between each zone-to-zone pair follow the same route. This technique often leads to unrealistic over and under loading on some links, so the concept of "capacity restraint" is often applied subsequently in an attempt to rationalise the loads.

Capacity restraint generally involves the systematic adjustment of travel times or costs on over-loaded links (and sometimes also on under loaded links) to reflect the levels of service (speeds) which would prevail under these conditions. New shortest paths are obtained as a result of each adjustment, and new "all-or-nothing" assignments carried out, until flows and capacities are adequately balanced throughout the network.

Other forms of capacity restraint exist - for example, adjustments may be made in the generation or modal split models in order to balance car parking demand and supply.

More sophisticated assignment techniques use some form of multiple routing, in which portions of the trips for each zone-to-zone pair are assigned to each of the paths through the network which approximate to the shortest path(27,28).

Another method for determining routes and assigning trips has been developed from probabilistic considerations. These so called stochastic models determine a route based on the trip maker's "opinion" of the route's travel resistance. This "opinion" is generated by Monte Carlo methods. Different routes from a given origin to a given destination can be found. This model lies between the capacity restraint and the multiple routing types.

II.3.6 Calculation of peak hour traffic

In developing models to predict the flows on a transportation system, considerable freedom exists for selecting the time period for which predictions are to be made. However, the tendency has been to concentrate on predicting total daily flows. This is based on the assumption that daily travel patterns are more predictable than hourly patterns, so that models developed from 24 hour data will be more reliable than those related to peak (or other) hour conditions. Consequently, although very important for the design of roads and public transport systems relatively little effort has been put into finding formulae for calculating peak hour traffic compared with the effort devoted to models for 24 hour flows.

Peak hour effects have different implications for public transport systems compared with private transport, and different calculation methods are generally used. On the public transport system, peak conditions are often estimated using the work (and perhaps school) trip purpose because these journeys cause the peak loads on public transport. On the road system, the usual method adopted is to take the peak hour flows as being some percentage (say 10 per cent, though this may be varied according to location and type of route) of the 24 hour traffic produced by the assignment models. Since there are no exact rules for selecting the percentage figure, there is considerable latitude for error in estimating peak hour flows. A second, and equally error prone stage involves converting the peak hour flows into the directional flows needed for system design.

A refinement of the above method utilises percentage figures which differ for each trip purpose. In this way, a number of "peak hour trip matrices" are obtained from the distribution model, summed and converted into peak hour traffic on the transportation system by means of the assignment model.

The special problems of overloaded networks and resulting speed/flow relations may call for specific peak hour models. Since the peak hour is normally associated with the journey to work, an assignment model which uses only the home to work trip purpose provides a means for directly calculating peak hour flows. Experience in Copenhagen and Stockholm suggests that total traffic in the peak hours is represented by a little less than 50 per cent of the total home to work traffic. However, this percentage is likely to vary considerably with the size of the home to work flow.

II.3.7 Appraisal of four-step models*

In Section II.2.3 on theory it was noted that the four-step model is a special case of the General Share Model. In conventional use:

a) The trip generation is a function of only activity system variable $\underline{A}$ (income, car ownership, etc.) and not of any level of service variables $\underline{L}$. Thus, it is implicitly assumed that no possible change in travel times, fares or parking charges, etc., will affect the number of trips made per person. This assumption is clearly unrealistic, especially for non-work trips. This partly explains the overall increases in flows observed when new roads, etc., are built - increases which conventional four-step models have failed to predict.

b) There is no direct opportunity for accounting for changes in fares, parking charges, etc., in the distribution model function.

c) It is only at the assignment stage that supply functions are introduced in the four-step models in an attempt to relate supply and demand.

By a process of successive iterations these effects can be allowed for, but this is cumbersome and expensive. The four-step model system is the most widely used transportation systems analysis method. It has been applied in over 200 cities in the United States and in many other cities around the world. The development and acceptance of the approach over the last fifteen years is a major accomplishment; it is one of the first large-scale applications of modern systems analysis techniques to problems of public sector decision-making.

However, because of the limitations mentioned earlier, it is useful to examine the four-step process critically from the perspective of equilibrium theory.

The objective of the equilibrium calculations is to predict the flows V_{klmp} i.e. the volume of trips going from zone k to zone l by mode m and path p. In the four-step models, equilibrium calculations are structured into a sequence of four steps; this amounts to estimating V_{klmp} in a series of "successive approximations" first V_k, then V_{kl}, then V_{klm}, and finally V_{klmp}.

It seems obvious that the following conditions should be met by any set of demand models and equilibrium calculating procedures:

1) The level of service attributes used should be as complete as necessary to adequately predict travel behaviour. For example, time reliability, number of transfers, privacy, etc., should be included if empirical evidence indicates these are important.

2) Level of service should enter into every step, including trip generation (unless an analysis of the data indicates in a specific situation that trip generation is, in fact, independent of level of service for all market segments over the full range of levels of service to be studied).

---

* See Appendix 1 for definition of variables used here.

3) The same attributes of service level should influence each step (unless the data indicates otherwise). For example, public transport fares, car parking charges, walking distances and service frequencies should influence not only modal split but also assignment, generation and distribution.

4) The process should calculate a valid "equilibrium" of supply and demand; the same values of each of the level of service variables should influence each step. For example, the travel times that are used as inputs for modal split, distribution, and even generation, should be the same as those which are output as results from assignmnnt. If necessary, iteration from assignment back to generation, distribution, etc., should be done to obtain this equilibrium.

5) The levels of service of every mode should influence demand. Congestion on highway or public transport networks, limited capacity (e.g. car parks), fares, etc., of each mode should (in general) affect not only its own demand but also the demand for other modes, at all steps (generation, distribution, modal split and assignment). That is, there should be provision for explicit cross-elasticities.

6) The several demand functions for each step should be internally consistent.

7) The estimation procedures should be statistically valid and reproducible.

Careful examination of the four-step model indicates it violates each of these conditions. As a consequence, serious questions can be raised about the biases and limitations of the flow predictions resulting from use of the four-step model. Whilst the model does have important advantages, these do not outweigh its very serious liabilities.

II.4 Demand Models

II.4.1 Demand/supply equilibrium models

As pointed out in section II.2.2 demand models estimate the equilibrium flows in a direct single step instead of the sequence of steps as in the indirect approach. Hence, the functions of generation, distribution, and modal split (and potentially, route choice) are combined in a single process.

The first such models were developed for forecasting intercity passenger travel(11-15) for the Northeast Corridor Project of the United States Department of Transportation, beginning with the Kraft-SARC model, and followed by the work of McLynn, Baumol, Quandt et al (12-15). Later work extended these models to urban travel(16,17). These types of explicit demand models were first used for transportation network analysis in the simulation studies for the Northeast Corridor Project. The Harloff model STADT to predict an optimal arrangement of land use is also a demand/supply equilibrium. models(35). The 'DODOTRANS' system of computer models(29) provides a variety of means of computing equilibrium flows in networks utilising various levels of service attributes and with the choice of demand models.

Other examples of equilibrium models in the general class are the models used at the TRRL(18,30,31,88) to study urban traffic restraint (using the RRLTAP suite) and for strategic studies (CRISTAL)(19). In both the RRLTAP and the CRISTAL approaches the criterion for decision to travel or for route choice is generalised cost of travel. The demand for travel between two points in the system is balanced against this cost using a demand/cost relationship which incorporates an elasticity for the travellers concerned.

The basic iteration loop is shown in figure 5. This concept embodies the essential reaction between supply and demand, permits a high degree of internal consistency in the models since a form of generation and distribution, also assignment and evaluation are on a common framework, and this greatly facilitates the study of the influence of changes in user costs on travel behaviour. There is some difficulty in the choice of appropriate (behavioural) time valuations and user elasticities but the effect of varying these factors can be studied.

The basic sequence of operations represented in a typical RRLTAP model is shown in figure 6.

It has been demonstrated that convergent iteration can be assured and the equilibrium points are stable, unique and well defined. It is also noteworthy that a method of assignment to congested networks has been adopted which does not involve any compromise with the basic benefit evaluation philosophy. CRISTAL(19) follows the principle of an equilibrium model in which an elastic travel demand (adjusted according to car availability ratios), is balanced against a function describing the generalised cost of travel on a network (see figure 7). Five modes (car, bus, rail, taxi and goods) of transport are included and their characteristics are modelled in considerable detail, modal split being dependent upon a chosen cross-elasticity in an inverse power formulation. Peak and off-peak hours are modelled separately and the effective network used has 400 nodes each of which is regarded as a possible origin or destination point. The simplifying feature which makes this model suitable for strategic studies is the adoption of a circularly symmetrical network with 20 rings and 20 radials; this can be described completely by the data for 19 radial links plus 20 ring links. Each of these links consists of an ordinary road and, when appropriate, a motorway and/or a railway (plus access and walking links). The model has been calibrated against data for London.

Outputs include the consumers surpluses of travellers and for road goods vehicles, the operating costs and revenues of public transport operators, taxation revenues, traffic flows (in aggregate and by link), O-D matrices, journey times, etc.

Experience with this model has demonstrated that it is well suited to strategic studies in which a wide variety of alternatives need to be studied without too much detailed information on each. The importance of traffic generation and repression mechanisms which are built into the model should again be emphasised in view of the importance of this feature in strategic work.

It should be mentioned that the French ASTARTE model seems to fall into the class of elastic demand/supply equilibrium models but details were not presented to the Group.

Figure 5

BASIC ITERATION LOOP

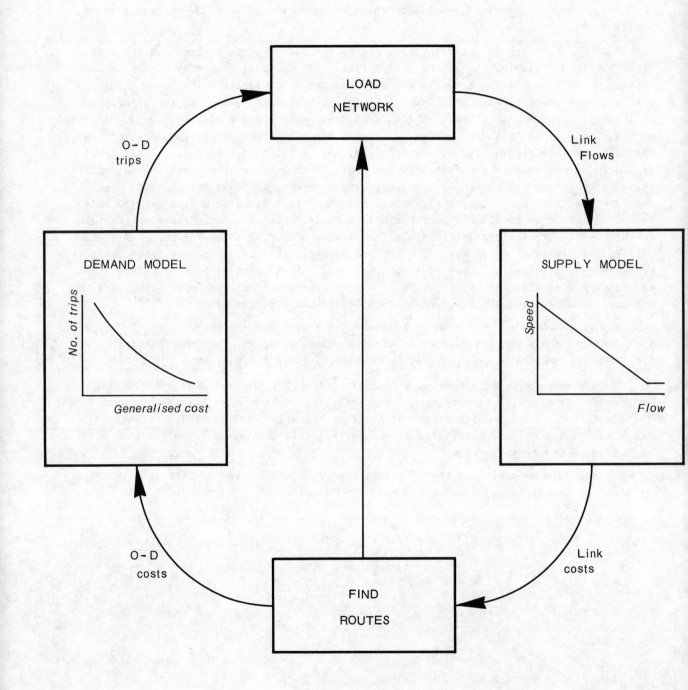

Figure 6

NETWORK MODEL FOR PARKING AND PRICING PROBLEMS

BASIC DATA

1. A set of roads forming the network

2. A set of points (A.B.C.) where all trips start and end

3. Number of trips between each pair of points from (A.B.C.)

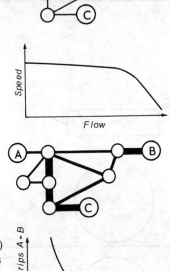

4. Measured relationship between traffic flow and speed for each road

STAGE ONE : Simulation of the current state

1. A pattern of flow over the network is found :
   All trips use a route of minimum (time and money) cost.
   and flows agree with the speed-flow behaviour of the roads.

2. There is a relationship between the total cost (time and money) of a trip A — B, and the number of trips made : step one fixes a point on this curve.

STAGE TWO

A specitic set of pricing charges is applied to the network. These charges may be either parking or road pricing charges, or both.

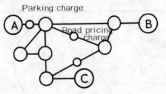

STAGE THREE : Simulation with charges

The new pattern of flows is found so that the corresponding number of trips agree with the altered cost of each trip

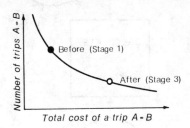

STAGE FOUR : Evaluation

1. Benefits to travellers
2. Changes in revenue collected
3. Changes in operating costs can all be obtained, and used to obtain a single figure of benefit

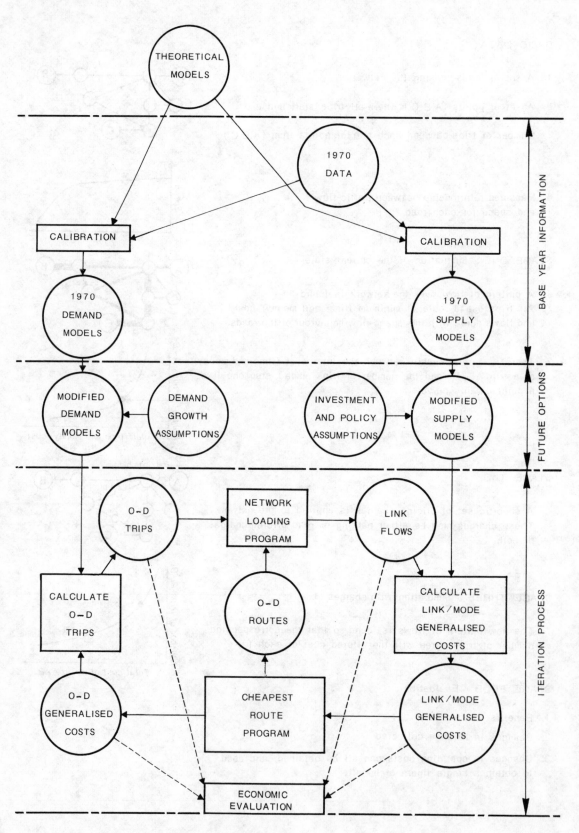

Figure 7

STRUCTURE OF CRISTAL MODEL

THEORETICAL MODELS

1970 DATA

CALIBRATION

CALIBRATION

1970 DEMAND MODELS

1970 SUPPLY MODELS

BASE YEAR INFORMATION

MODIFIED DEMAND MODELS

DEMAND GROWTH ASSUMPTIONS

INVESTMENT AND POLICY ASSUMPTIONS

MODIFIED SUPPLY MODELS

FUTURE OPTIONS

O–D TRIPS

NETWORK LOADING PROGRAM

LINK FLOWS

CALCULATE O–D TRIPS

O–D ROUTES

CALCULATE LINK/MODE GENERALISED COSTS

O–D GENERALISED COSTS

CHEAPEST ROUTE PROGRAM

LINK/MODE GENERALISED COSTS

ECONOMIC EVALUATION

ITERATION PROCESS

II.4.2 Explicit demand models

Since the four-step model as presently applied has serious limitations, many researchers have focused attention on developing explicit(11-17) models - i.e., models which can be written down as a single equation. Examples of these are given in Appendix 1.

A wide variety of forms of function are possible as well as choices of variables, of specific numerical values for direct and cross-elasticities and of forms of the "generalised cost" function.

Alternative forms of the "generalised cost," can be used, e.g.,an exponential transform of this(33), a friction factor transformation of travel time, etc. (see 11-15).

Recent research has refined the form of models. In early work for the Northeast Corridor project in the United States, the Kraft-SARC model was developed for intercity passenger traffic. This model, as illustrated in Appendix 1 features (1) incorporation of several activity system variables; (2) incorporation of several level of service variables (time and cost); and (3) incorporation of "cross-elasticity" effects, i.e., the demand for travel by mode m is affected not only by its own level of service but also by the service offered by competing modes n. Other models developed later - e.g., the McLynn and Baumol-Quandt - have similar properties, although the cross-elasticity effects were not expressed so clearly in the model structure(14).

II.5   Urban Dynamic Development Models

The field of strategic traffic models involves close contact with land use planning and dynamic models of the resulting interaction are currently being developed.

The main goal of an urban dynamic development model is to collect and assemble all aspects of town development, to evaluate their interrelationships in an optimising process and to quantify their implications. The data requirements imply that they cannot be regarded as simplified models.

The Polis model(34) is still being developed. It will eventually describe urban development with regard to traffic and land use; furthermore it will consider economic, social, cultural and political implications as well as environmental aspects. The model consists of several submodels with sequential runs as well as feed-back-loops. The model becomes dynamic by introducing time intervals, which may be varied by length and number. The traffic system is taken as the independent variable while land use is considered to be the dependent variable. The model structure is contrary to the normal transportation planning algorithm. The supply by a traffic system is considered to be the main impulse to urban development. The simulation makes use of traffic generation-, attraction-, distribution-modal-split- and assignment-models.

The Stadt-model(35) attaches the various types of land use to areas in a town and to the regions attached to it by a relation between land use and essential supply services. The goal-function in the standards of linear programming tries to minimize overall costs taking due account of traffic loading and any constraints. The essential supply of the types of land use must be achieved for transport systems with restrained capacities. The capacity restraints of these systems are the constraints for the allocation of land-use units. The computed model output is the amount of traffic by mode on the system in addition to data on land-use in each cell in the planning area. The system of numerous constraint-equations is able to formulate the conditions of the real world, of social standards and goals and of a development without contradiction.

Another group of models is based on the Lowry-model. One of these is the Swiss ORL-MOD 1(36). The elements of ORL-MOD 1 are nearly the same as in the Lowry-model.

Around this, distributing the activities of land-use, several sectoral single-purpose models have been grouped to assemble data, to compile them and to prepare the outputs.

A third group consists of models structured by methods of systems analysis. This is the most promising group, though they are in the moment only global models. The basis of these models is the Forrester model, mentioned in "Urban Dynamics"(37). In short the model simulates the economic development and growth of population of a town over a period, without changing the borders of the town. It simulates the impacts of town-planning programmes for the economic redevelopment of the town over a 50 years' period. As there is no division of the area of the town into small (discrete) cells the model only simu-lates global developments. The model consists of several positive and negative feed-back-loops, concerning work places and industry, supply by dwelling units and population. The whole model system is a closed system with programmed cycles and their impacts.

The Besi-model(38) is in its model structure very similar to the Forrester model. It consists of several sub-models:
- managing the goal system of town development,
- handling data for an informations system,
- operating for a strategic planning phase,
- operating for a tactical decision phase,
- describing the operating system.

The Besi-model has been conceived as a community management and information system setting out the structural changes in a large town according to their economic impacts. The whole set of submodels is not yet developed.

In general no urban dynamic development model may be regarded as an ideal one. Methods of systems analysis seem to be a promising way in developing better models, which make it possible to take into consideration a great variety of alternative concepts at reasonable costs in a reasonable time.

II.6 Appraisal of Implications for Simplified Models

The above discussion has a number of practical implications for the development of simplified urban travel forecasting models.

i) It can be proved that any explicit demand model is a special case of the general share model (GSM): that any sequential implicit demand model (which meets certain "internal consistency" conditions) is also a special case of the GSM; and that as a consequence, every demand model can be used in either its explicit or its sequential implicit forms(7). Therefore, the developer of simplified models is free to choose whether he wants to use a single-step (i.e., direct) approach to computing equilibrium with an explicit demand model; or a multi-step (i.e., indirect) approach with a sequential implicit demand model.

ii) In particular, the conventional four-step approach is only one of many possible approaches. Because the four-step model has serious limitations, other approaches should be explored actively.

iii) It is practical to calibrate various forms of explicit demand models. Empiri-cal results already exist(11-17).

iv) Particular attention should be given to the question of which and how many level of service and activity system variables should be included in any model. (The limitations in the generation, distribution, and assignment phases of the four-step model probably introduce significant biases).

v) Thus it has been demonstrated that there is a systematic theoretical link between explicit demand models and more conventional transportation models. This has indicated the importance of the explicit demand models in avoiding large systematic errors and in opening the way for reducing the amount of data which has to be handled to describe real situations effectively.

vi) Despite the foregoing conclusions drawn from work on the theory of transport models, the majority of simplified models at present in use still follow the conventional four-step approach. Internal consistency is a vital feature of models which are to be used for the strategic phase of urban planning in which possible generation or suppression of traffic will play an important part and this is not usually present in four-step models. Equilibrium models are especially suitable for use in simplified traffic models since it is possible to demonstrate the accuracy of solutions derived. This feature is largely lacking in the overall results obtained by four-step processes. The attraction of the four-step process is that items like traffic generation can be separated out and related to physical proposals but care must be exercised to avoid the danger of thereby introducing an unrealistic bias. It should be pointed out however that equilibrium models can also provide this facility.

vii) The CRISTAL model is an example of simplification in the network in order to avoid compromise on the equilibrium formulation in the model. This suggests that simpler models do not necessarily conflict with the idea of correctly representing demand/supply interaction.

viii) There is no doubt that a lot can still be done using versions of the conventional four-step model but work on the theory and practice of a new generation of explicit demand models and methods of improving data inputs (see Chapter IV) promises to offer important improvements in the effectiveness of simple models.

III

COMPONENTS OF STRATEGIC MODELS

The following sections contains a review of current work with regard to model simplification, a first section containing an overview and the latter sections studying the issues that arise in connection with particular aspects of the transport planning process.

III.1 Current Work on Model Simplification*

III.1.1 Introduction

In all participating O.E.C.D. countries, considerable use has been made of conventional (four-step) transportation planning models of the type outlined in Section II.3. A number of internationally used suites of transportation analysis computer programmes exist (e.g. United States Department of Transportation packages). In addition, some countries have developed their own suites to meet national and local needs (e.g. in Germany 4-6,10,24-28,40-45). These programme packages, which are built around conventional model techniques, are not considered in this section. Instead, a description is given of ways in which Member countries have simplified transportation planning models either by building on the conventional processes or by developing new methods. Section V.2 describes ways through which further simplifications could be introduced.

In most participating countries simplifications to conventional models have been introduced in the data collection stages, in the way in which the various travel purposes or modes are treated, in the land use groupings, in the degree of accuracy acceptable for travel predictions, in the details of the mathematical formulae and in the study procedure itself. Some countries also report simplifications through new methods of handling networks, new forms of the mathematical formulae built into models, and developments in computer procedures(24).

Despite these simplifications, the resulting models are often quite complicated when attempting to accommodate a large number of trip purposes, travel modes and zone numbers in their generation and distribution stages. As mentioned in Section II.3.1, this is probably due to the ease with which computer based analysis can handle the resultant mass of data. One could almost say that current urban transportation models owe their complexity as much to the size and availability of computers as to difficulties inherent in modelling human behaviour.

III.1.2 Simplification of conventional models

Five main types of simplification within the methodology of conventional (four-step) models have been used in Member countries.

In the first method(46a), simplification is achieved through the restriction of the models to two purposes (home/work and home/other). In the case of home/work travel, calculations are made with a gravity model using numbers of employed residents and employments per zone. Regression analysis based on car ownership level is used for estimating home/other travel. Tests indicate that the models reproduce survey data to within 15 per cent.

---

* The national reports(46) prepared by Members of the Group can be obtained on request to O.E.C.D.

In the second method, simplification is achieved by adopting a regional rather than detailed level of analysis(46b). This has permitted simplification in the data collection stages, with travel models being developed on the basis of a 1 per cent home interview sample size. A category analysis trip generation model permitted the reduction in sample size. Modal split was incorporated in the generation model. The validity of the model process was checked by comparing synthesised and observed flows across selected cordons; a calibration accuracy of within 9 per cent was indicated(47).

The third method of simplification concentrates on reducing computer analysis time by adopting an "in-core" set of programmes, a "once-through" tree building and loading algorithm and a doubly constrained gravity model for rapid convergence. This has been used in the IMPACT model(46c) which can handle up to two modes, 150 zones, 500 nodes and 2,000 links. A generalised cost function is used to reflect travel impedance with a simplified method of estimating intrazonal costs. Applications of the model have included examinations of large investment alternatives. It has been found possible to aggregate zones for this purpose of analysis, with meaningful results being obtained with aggregation from 362 to 32 zones.

Another simplification consists of a modification to the gravity model formula so that a future trip matrix can be predicted from an existing matrix(46d). This modification uses only the future and existing zonal generation and attraction values (based on expected land use alterations) and the existing trip matrix.

Finally, simplification in the overall analysis effort has been reported from Spain and Sweden where the models developed in one city have been applied directly in another city without having to carry out large O/D surveys. In Sweden a test procedure is used to confirm that travel data do apply in the new city. A forecast is made for the present situation which is checked against measured flows and calibrated until predicted and measured flows agree to a satisfactory degree.

III.1.3 Simplification of modelling procedure

Three types of simplified model procedure are reported from Member countries.

The first makes use of information collected in a number of "conventional" travel studies(48). Average generation factors are calculated and used to predict the private travel generated by suburban residential areas. Various formulae are presented. These can be used to provide a rapid estimate of traffic conditions, with more complicated (conventional) models used to determine the final network loads.

The second type(46e) involves restricting the process to one category of trip and combining the trip generation and modal split stages. In addition, it involves dividing the forecasting stage into two steps. The first step uses growth factor techniques to predict the growth in travel arising as a result of increased affluence, etc., among the existing inhabitants of the city. The second step estimates the growth in travel due solely to new population, new developments, etc. A number of simplifying methods are suggested for completing this second step. All dispense with home interview surveys and instead rely on traffic counts, on screen lines or on average trip generation values from other surveys. (See section V.2.5 and Appendix 3.)

The third type(46f) dispenses with comprehensive O/D surveys and instead makes use of the averages of travel data collected by previous studies. It divides the study area into a small number of zones for each of three classifications. The criteria for zone boundaries are different for each classification, so zones may coincide or overlap. For each zone, normal information (such as population, employment and land use) is collected, along with much information of a qualitative nature. These along with the average travel data and models from other studies, produce synthetic travel patterns which can be checked for validity by comparison with observed traffic flows at selected points on the network. Computer coding and analysis is not required, since a simple zone and network system and assignment procedure are used. There is close co-operation between the land-use planners and transportation planners at all stages of the analysis and evaluation. Because of the reduced amount of data collection and analysis, more effort can be devoted to examining and evaluating different land use network alternatives.

III.1.4 Simplification of external aspects of models

The main features of this type of simplification lie in having models with clear logic so that non-technical decision makers can understand the relationship between input and output. They can then appreciate the close link between land use and transport infrastructural needs, and understand better the problems of the city centre where most density and network conflicts arise(6). These simple, easy to operate models have been developed in France. They require few input data, produce clear output information, give a rapid turnaround and can be easily and simply explained. The models consist of a conventional generation-distribution sequence, with complementary stages for modal split, assignment to predetermined routes and analysis of city centre parking. The models are based on the computer packages FABER and PARK(49), which have been used in France to assist in the three levels of planning in all towns over 10,000 population. Validity has been tested using the results from home interviews in 17 comparable French towns. Information from these 17 surveys has been assembled into a data bank. On comparison with the data bank, the models were found to reproduce average survey trips with an overall deviation of 10-19 per cent. Smaller deviations (10 per cent - 15 per cent) are obtained when the model gives different weightings to trips depending on whether they originate in the city centre or suburbs.

III.1.5 New modelling techniques

Network handling

Three methods of simplifying models through improved network handling are reported from Member countries.

In the first method, a model has been developed which generated networks for investigation during the planning of the Sevilla underground system(46g).

The assumption of a circular symmetrical network is basic to the technique developed by the Transport and Road Research Laboratory and used in their CRISTAL model (see section II.4.1). As a result, computer analysis time is greatly reduced, and many strategic alternatives can be examined.

Another simplified model developed(89) involved the representation of the study area by its "average main road". The model has been used to determine optimum bus size, and calculate community savings arising from a transfer from cars to buses. In the model, the average speed on the network is calculated from observed speeds and speed/flow relationships at various points. Speeds are modified to allow for car parking and bus walk/wait times. Travel time is then converted to journey costs per kilometre of network per hour, and the procedure repeated for various car restraint and public transport size options.

Equilibrium methods

New mathematical techniques are being investigated which could aid the development of simplified models. Some of these techniques are based on the concept of the General Share Model (GSM). Starting with a mathematical expression for the total amount of travel in the study area, the GSM successively splits this total until it has identified the fraction of this total which is made on a particular path by a particular mode between a particular pair of zones (see also Appendix 1).

A special case of the GSM is the Explicit Demand Model (explicit, in that volume appears only on one side of the equation). This provides a "single step" method for computing equilibrium in the transportation network. In the more general case, the GSM can be written in an implicit form (implicit, in that volume appears on both sides of the equation). A special case of the implicit form is the sequential implicit model, in which equilibrium is calculated in a series of steps analagous to generation/distribution/modal split/assignment.

Calibration methods

Previously developed demand models have been aggregate deterministic models in that they describe the behaviour of groups of people, indicating the number of people who travel under specified level of service conditions. Recent work on disaggregate stochastic models deals with the travel behaviour of individuals and give explicitly the probability of the individual's making a specific choice (4,5,20,21,22,75,76,77). These disaggregate models allow much greater efficiency in the use of small samples of the order of 1,000 observations for calibration. Once developed, disaggregate models can be transformed into equivalent aggregate ones.

Computer aspects

Problem oriented computer languages which have been developed for many analysis techniques have only recently been applied in the field of urban transportation. They simplify user communication with the computer when defining or operating models.

The Massachusetts Institute of Technology have developed one such language for transportation analysis, i.e. DODOTRANS(29). In outline form, DODOTRANS starts with specified options (in technologies, networks, operating policies and activity systems). The consequence (volumes, levels of service, resource requirements and activity system changes) arising from these options are predicted (by one of four demand models), and the impacts (to users, operators, government, and functions) are evaluated. A search for transportation alternatives may then be made either explicitly or by mathematical optimisation. If necessary, some of the original options can be altered and the procedure repeated. DODOTRANS has been used to analyse experimentally the Northeast corridor region in the United States and for studies of new public transport systems such as personal rapid transit.

Following experience with the FABER models(49), which are being replaced by the ATLANTE model(78), a special computer language based on macro-instructions has been developed in France. Each macro-instruction initiates a predefined sequence of computations, for example, trip generation, modal split, calculation of indicators for planning criteria. The instructions can call several independent programme modules and can, for example, utilise gravity or intervening opportunity models for trip distribution. Provision is made for further development through the addition of future programme modules.

Similarly, the RRLTAP system(90) used in the United Kingdom provides a modular suite of transportation functions which can be assembled into a variety of transport models with maximum flexibility and ease of handling. It should be mentioned, however, that this is essentially a methodological research tool and, although completely documented and available to other users, has so far been used mainly at TRRL.

One of the main aspects of these computer procedures is the way in which they have simplified communication between the transportation analyst and the computer. Other ways in which simplifications can be achieved include easier access to the computer through the use of remote terminals (and the development of transportation planning computer programmes to facilitate this), simplifications in the instructions needed to operate the suites of programmes and improvements to the programmes to give greater clarity of output including output on graphical display devices(50) and use of analog computing devices(51).

III.2 Changes in Land Use

The main, and often the only, aim of early urban transport planning studies was to satisfy traffic demand estimated on the basis of the underlying urban development assumptions and objectives. Thus, the traffic models used were simply aiming at estimating, with the highest degree of accuracy possible, the amount of traffic in the target year.

Almost all traffic models presently available were developed along the same line. However, the relationship between transport and urban development is not one-way; transport conditions have a considerable effect on the development of an urban area (see also section II.5). Results of traffic studies are therefore being used more and more to modify and adjust land use variables instead of considering them as being intangible basic assumptions likely to prove unrealistic. It is thus possible to obtain dynamic interaction models of transport and urban planning which reflect the interdependency of these two sectors.

In view of the modification of land use variables the two following aspects related to the application of models are successively considered:

i) the relationship between transport planning and land use,

ii) the implications for traffic models and possibilities for simplification.

III.2.1 The relationship between transport planning and land use

Application of traffic models to land use problems may be put under three headings:

- long term static coherence of transport network and urban development;
- dynamic interaction of accessibility and urban development;
- control of land use by regulation and pricing

Long term coherence of transport network and town planning objectives

This phase of research to achieve long term coherence is important in that it is often the first and sometimes the only aspect to be approached in the practical planning process. The introduction of land use characteristics is necessary because they determine the amount of traffic generated and consequently the transport network to be provided. However, the technical and economic difficulties they raise concerning transport facilities may very often be considerable and the initial urban planning objectives may therefore have to be re-examined. Experience has shown that the general plan and objectives finally adopted for urban development almost always stray away from the initial concept and this result of applying traffic models is just as important as the development of the transport system.

Studies have brought out two particularly critical problems that could justify a revision of urban development assumptions: the problem of increasing density in the centre and that of establishing a public transport system with its own right of way.

The problem of increasing density in the centre. Elected representatives and town planners usually favour increasing the density of employment (especially tertiary) in the centre and also, in some cases, that of the population.

The function of the traffic model is to show the cost (in terms of transport infrastructure) of pursuing such a policy.

To contribute to this, models should faithfully represent the traffic variations as a result of varying land uses in the centre. It will not be sufficient, in particular, to know the overall number, per zone, of tertiary sector occupations: indeed the traffic generating pull of such occupations varies considerably, some of them being in offices closed to the public and other commercial occupations attract a great deal of traffic. In one centre, 62 per cent of the trips were attracted by 34 per cent of the occupations.

Generally speaking, application of the model leads up to the following (policy) options:

- either reduce the growth of central tertiary occupations;
- or step up provision of new roadways and parking places;
- or, in fairly important towns, create "segregated" public transport systems.

Whatever the choice or combination of choices, it will have repercussions on land use which must be clearly expressed.

Establishment of a public transport system with its own right of way. The success of such a project closely depends on the policy in regard to land occupation near the stations. Patronage of public transport does indeed depend to a very great extent on the terminal time and walking distances that have to be covered on foot.

It will be generally necessary to:

- concentrate most of the new central area tertiary occupations within 300 m of the central area stops.

- concentrate within 500 m of peripheral stops the largest possible proportion of new collective housing.

Such projects should therefore be assessed by means of a public transport patronage model sensitive to terminal time and walking distance.

Dynamic interaction of accessibility and urban development

A new public transport infrastructure (own right-of-way or metro) can very considerably alter the accessibility map (see Appendix 4) of a built-up area and it will also directly and powerfully affect spontaneous urban development.

Long term town planning cannot therefore merely seek to maintain static coherence between development and public transport; it will also have to check, at each stage of transport development, that the stimulus to urban development is in keeping with the final objective; it will thus be possible to review policy and control of land use(41).

Mutual interaction models of accessibility and town planning have been used in France.

The concept of these models is based on the measurement of overall satisfaction of residents, employed or not, in terms of freedom of choice. Accessibility which combines both the various choices offered by the city and the possibility of access reflects better than other variables the overall satisfaction obtained.

Appendix 4 shows that it is possible to evaluate benefits to the citizens provided by the city in general including the transport system. It is also possible to
determine urban development patterns if it is assumed that each person tries to obtain
the best benefits.

The aim is not necessarily to create an exact model of urban development. It
seems, however, that the use of an accessibility model improves the validity of the basic
land use assumptions and that the model could become a basic tool for urban planners.

Control of land use by regulation and pricing

Traffic models can have several applications in this field:

- regulation of the number of parking places in new buildings: construction
  of parking places to be encouraged or locally limited according to circumstances;

- a differential tax according to central area land use.

It has been calculated that each new office job in Central Paris costs the community 40,000 F (mainly in investment and operation of the transport systems).

Transport models may thus be used for real estate market guidance in step with
"real costs".

III.2.2 The implications for traffic models and possibilities for simplification

The first objective is to obtain a rough estimate of generated traffic on the basis
of given land use assumptions.

For optimum refinement, a considerable number of variables must be taken into
account, the number soon becoming too large for practical purposes. In particular, the
following factors should be taken into account for each zone:

- residents, by household size, income level and degree of motorisation;

- employed residents, by category of occupation (this will easily amount to
  10 to 20 categories, i.e., shop, office, etc.);

- point generators: large stores, hospitals, schools.

The need for simplification is evident. In France, simplified strategic generation models use the following variables (applied to 30 to 50 zones):

- total number of residents
- total number of employed residents
- total number of occupations
- proportion of tertiary occupations

These models have been tested by comparison with the data gained from household
survey interviews. The weighted error affecting traffic generation is of the order of
10 to 20 per cent (variable according to motivations) assuming that the overall number
of trips in the urban area is accurately known.

Obviously, part of the error is the result of simplification. For example, the
models do not take into account the relative distribution of tertiary occupations, i.e.,
the breakdown between really attractive employments (stores, offices open to the public)
and places of employment with low degree of attraction (offices closed to the public).

It would appear that the only solution would be to differentiate tertiary occupations into more homogeneous categories, this may well lead to resorting to more complex
approaches, which would defeat the very object of simplification.

However, accuracy can be improved without adding to the complexity of the land use parameters. It has been found that the relative distribution of tertiary occupations is fairly constant in the centres, the first and the second rings of French towns, hence the concept of treating land use data differently according to the type of zone (centre, first ring and second ring) has led to the development of a set of correction coefficients to be applied to traffic generation and attraction according to the type of zone and trip motivation.

It has been shown that this approach results in a marked gain in accuracy, especially as regards trips related to the central zones where transport problems are the most severe.

It will be noted that this increase in accuracy is obtained without adding to the weight of required data, the only supplementary information being the type of zone (centre, first and second ring).

The second objective worthy of mentioning relates to the dynamic interaction of accessibility and urban planning. It assumes that for each zone the accessibility can be calculated.

In this connection, the problem concerns not so much the land use data but, rather, the measurement of trip times. Trip times, especially when the transport networks are congested, can be assessed only after implementation of traffic distribution models with capacity restraint, which can not be considered as simplified models.

A degree of simplification can be attempted by substituting, for the trip times, point to point distances, in the accessibility relation (see Appendix 4). In that case, simplified models of the type described may serve the desired purpose though at the cost of accuracy (especially in the presence of traffic congestion).

III.3 Public Transport Aspects

Many problems in the field of public transport may be tackled with the help of simplified models (optimum positioning of bus stops, organisation of connections, priority at crossings, etc.).

If it is confined to long term planning, the main contribution to be expected from a model of public transport demand will be evidence that any improvement of the situation will not automatically follow a stroke of a planner's pen and that it requires a very considerable effort of improvement in supply.

Such models should therefore essentially establish and quantify the link between the supply of public transport and the demand for it. Here again, it is essential that such models should be as logically simple, convincing and instructive as possible in their choice of explanatory variables. This is essential to convince policy makers and the administration of the project and to dispel any impression that their decisions will be arbitrarily dictated by an incomprehensible model.

At first sight, the demand for public transport seems to be governed by a large number of factors including:

- choice (possibility or not of using a car)
- trip purpose (in town, suburb to centre, suburb to suburb)
- frequency (waiting time)
- operating speed
- comfort
- number of changes

- initial and terminal distances on foot (concentration of dwellings and employment around stops)
- reliability of time-table (respect of design frequency)
- fares
- own right of way or shared use of roadway
- parking facility in town (charges, prohibition and repression of parking,
  control policy, time required to find a place)
- congestion level of roadway.

These variables are not all independent of each other.

It is, of course, possible to design a model which will take every one of them
into account, but it would be too unwieldy for strategic planning.

Another approach must therefore be used by arranging the factors mentioned in
order of priority and by including only those strictly required to obtain the required
degree of precision (it is assumed that the factors left out are partly reflected in
those that have been included).

Particular examples of simplified models are considered below.

III.3.1 Generation models by mode of transport

On the basis of total traffic generated, a separation between the different modes
is made by means of an overall evaluation of the service provided to each zone by each
mode.

These models do not indicate the relationship between the transport network (quality of service of each public transport line) and public transport patronage and this
disadvantage cannot be avoided. In fact the quality of service can only be taken into
account once traffic has been distributed to the different modes of transport.

III.3.2 Models based on tables of modal split

These models are based on a set of tables giving the level of public transport
use according to zone of origin and destination (centre, first or second ring).

The tables will generally have to be split up according to trip purpose and whether
or not public transport has its own right-of-way. Car ownership which has a very important influence on individual choice does not need to appear as such if it is assumed to
be more or less equal throughout the centre, and throughout each of the first and second
zone "rings". It will be sufficient to take it generally into account in drawing up
the table.

The following is an example of a table used for 1985 in a built-up area of which
the present population is 170,000.

Table 2

Level of Public Transport Patronage

(% of trips by vehicle)

Type of trip	Home-work	Home-other purposes	Secondary
In town centre	5	5	5
Town centre - first ring	13	12	6
Town centre - second ring	20	14	5
Peripheric	15	13	5

Although such a method is very workable, it is rather unrefined and uses levels that are somewhat arbitrary. It can be used for an initial approach, but it cannot be said that it satisfactorily quantifies the link between the quality of service and the load factor or patronage of public transport.

This drawback however may be partly overcome by refining the tables, i.e. by sub-dividing the area considered and identifying the zones affected by public transport (in general a corridor of 500 m width either side of the public transport line). This will take into account public transport right-of-way effects and interchange penalties.

The following is a table used for Lyon, France in 1985 (population 1.7 million in 1985). This table presents public transport patronage for the zones affected along a route as a function of the quality of the mode (subway, bus, direct lines or with inter-change) and the type of zone (e.g. parking possibilities).

Table 3

Patronage of Public Transport for the zones of influence along a line

(% of trips by vehicle)

	Town centre-town centre		Outskirts to centre		Centre to outskirts		Outskirts to outskirts	
	Work	Other	Work	Other	Work	Other	Work	Other
Subway direct	85	70	80	50	70	40	50	35
Subway + subway	80	60	70	45	65	35	45	30
Subway + bus	75	50	60	35	55	30	40	25
Direct bus and bus + bus	70	35	40	25	35	25	30	15

III.3.3 Models comparing level of service of different modes

However refined they may be, the tables are still to some extent arbitrary and empirical and only imperfectly reflect many of the above-mentioned factors.

If a better assessment is required of the impact of different measures connected with public transport - and more particularly of the patronage to be expected on a new line or service having its own right-of-way - it will be necessary to go back one step in the priorities and compare the level of service offered to users by the different modes of transport (private car, public transport, two wheeled vehicles).

Systematic research was undertaken in France into the form of such a model and on the minimum number of variables to be taken into account. Basic data were obtained from house-hold interviews. This research has given the following results:

Zoning

A detailed link by link analysis calls first of all for zoning sufficiently refined to show whether dwellings and work destinations are accessible from the public transport considered (i.e. less than 500 m from a stop).

User categories

A distinction has to be made between two categories of users;

- those who have a car and usually choose between their car and public transport
- those who do not have a car and who have to choose between public transport and possibly a two wheeled vehicle.

Choice between private car and public transport

Trip purposes were split up under "home-compulsory" (work, school, daily shopping), "home-optional" and "secondary".

The explanatory factor to be determined was the relationship between travel time by private car and by public transport. This factor gave satisfactory results. However, it was necessary to introduce the following corrections with regard to travel time:

- private car: penalties reflecting parking difficulties

- public transport
  - a coefficient of attraction in the case of own right-of-way
  - a coefficient for interchange penalties

In order to simplify and avoid a generalised cost model, it is possible:

a) private car
   - to standardise the time penalties due to parking problems by only differentiating between the type of zone (CBD, centre, first ring, second ring)

b) public transport
   - to indicate that the penalties mainly concern short trips (waiting and walking times representing a considerable portion). The penalty coefficient is:

$$1 + \frac{15}{d} \quad \text{where d is the distance in km.}$$

Choice between private car and two wheeled vehicles

It is sufficient to correct the time per two wheeled vehicle by a coefficient representing the density of public transport lines (expressed by the logarithm of urban density).

III.3.4 Explicit demand models

The traveller's decision process in public transport mode choice is complex. Some current work(79,80) utilises explicit demand models estimated statistically from available data(17).

Essentially, these methods use the demand model in a "pivot point" approach based upon existing conditions. The elasticity E of public transport volume with respect to a change in some level of service attribute, such as schedule frequency, can be devised from the explicit demand model. Then, given the existing volume over a route V_o and existing frequency F_o, and the new frequency F_i, the new volume can be computed approximately as:

$$V_i - V_o = E \cdot \frac{(F_i - F_o)}{F_o} \cdot V_o$$

This method has been applied by hand for very quick and inexpensive appraisal of changes in public transport service.

III.3.5 Conclusions

The main conclusions are as follows:

1) A simple table of modal split will give a rough estimate of public transport patronage for strategic planning purposes. Such a table is not however very sensitive, except by means of arbitrary corrections to the main variables (improvement in frequency, establishment of own right-of-way, urban density around stops).

2) More complex models based on comparison of times and level of service of different modes give satisfactory results with curves that are independent of the built-up area considered - and which are therefore universal.

These models have the advantage of being immediately sensitive to the planning operation variables already mentioned.

3) It is however, possible to simplify those modal split models which involve the comparisons of generalised costs.

There are three possibilities:

a) the generalised trip costs could be approximately evaluated by using the distance as the crow flies. This means that one can avoid breaking down these trips, connection by connection, into their various components (walk, waiting for bus, time spent looking for a parking space, etc.)

b) detailed analysis of car journey times (which entails the use of elaborate models to simulate congestion) can be avoided if one uses "a priori" the average speed on the network.

To simplify even more, direct comparison between travel condition by car or by public transport can be avoided; one only needs a variable for the quality of service in public transport (e.g. the number of changes). The ultimate stage in simplification is the use of models based on tables of modal split.

c) one could also consider combining all trip purposes, or all types of users (whether car owners or not).

However, even for strategic planning purposes, such simplification may have drawbacks. If one combines all trip purposes it is then impossible to determine the effects of a parking policy on public transport users; these effects differ considerably, depending on whether the trip is connected with work or for another reason.

4) Explicit demand models can be used to estimate modal behaviour, using a pivot point approach.

44

In conclusion it seems almost impossible to develop a modal split model which is both simplified and adapted to all kinds of strategic planning objectives. However, at least unnecessary complications of modal split models should be avoided.

III.4 Modelling New Modes

The problem that arises in using four-step models to assess new modes is that although the models are calibrated to determine the values of the zonal and deterrence parameters, these are not strictly parameters at all, but undertermined multipliers, i.e., they are variables of the system and any change in system characteristics will cause the undetermined multipliers to change also. This is not considered to be too serious a problem when it is assumed that the changes that are made to the transport systems will have only marginal effects on general expenditure patterns and changes in the unde-termined multipliers will be minimal. The consistency that has been obtained in calibrated values of the deterrence parameter in different studies is seen as some justification for this assumption.

III.4.1 Definition of new mode

Care must be taken about the definition of a mode (see also 52). Two questions arise, firstly the decision of whether or not the "new mode" in question is an addition to the number of modes in the system or is simply the upgrading of an existing mode; secondly, if it is decided that the "new mode" does indeed alter the number of modes in the system, can this still be seen as a marginal change that will not affect the values of the undetermined multipliers? The answers to these questions are at present unclear. A current development is to introduce a distribution of zone-to-zone generalised costs for journeys by each of the modes. Travellers are then assigned to the least cost mode. It is hoped that this procedure will remove the problems outlined above.

III.4.2 Abstract mode

A major problem in assessing new modes of transportation is that of specifying the disutility of travel by the various modes. Early work concentrated on the Abstract Mode approach. A mode is seen merely as a means of achieving a transformation through time and space, the traveller being required to make certain inputs into the process (money, time, etc.). Alternative modes are seen as alternative potential transformations and the task for the traveller is to choose that mode which yields the minimum level of disutility. A mode is thus seen as one combination of a set of common characteristics that relate to all modes, and it is in this sense that the modes are considered abstract. With this approach the introduction of a new mode is no problem, all that happens is that an additional potential transformation is introduced. The traveller processes the new information in the same way that he did with the original set.

III.4.3 Practical application

Although this framework allows the problem of the introduction of a new mode to be handled at a conceptual level, there are problems with the practical application of such a model. The difficulties relate to the empirical specification of the disutility function; in general, the modal characteristics that are easily quantifiable are money costs, travel times, and walking and waiting times. These characteristics have proved insufficient in explaining modal choice satisfactorily, and additional explanatory variables have had to be introduced that are specific to individual modes. These variables allow for the effects of the non-measurable characteristics and thus destroy the "abstract" nature of the approach. This creates great problems when a new mode is introduced into the system; it is necessary to know the value of the mode-specific parameter before the patronage that the mode will attract can be assessed. With a new mode, all that can be done, given the current state-of-the-art, is to guess the values of such parameters from intuitive reasoning such as: new mode X appears to provide a service that is inferior to that of a car but superior to that of existing buses; hence, the mode-specific parameter for X will lie between the values for car and bus. Such a procedure is imperfect, but necessary, if estimates of the patronage of new modes are to be made. Work is needed in this area of specification of disutility functions to allow a more satisfactory method of estimating usage for new modes.

III.4.4 Generated traffic

Trip generation is another area in which the usual assessment procedures are less than satisfactory for new modes of transport. These procedures relate zonal trip productions and attractions to the gross demographic and socio-economic characteristics of the relevant zones. Hence, the level of trip making is invariant to the transport networks. This is one of the more worrying aspects of the normal four-step process; the total level of trip making as well as the distribution of trips and the mode of transport chosen, would be expected to depend upon the quality of the transport networks provided. The equilibrium models used at TRRL overcome this difficulty to some extent but there is a lack of data on the generative aspects of transportation systems, and research is needed to investigate this question. Also most transportation studies do not concern themselves with walk trips or bicycle trips, but concentrate on vehicular transport. Thus, there is no mechanism to allow for diversion of these trips to vehicles. Some novel modes of transport are specifically geared to act as distribution systems, e.g., moving pavements, automatic tracked vehicle systems, etc., and are likely to be competitors of the non-mechanised modes. Failure to consider the effects of such systems on non-mechanised modes could lead to serious underestimates of potential patronage, but at present there is a shortage of adequate data for this purpose.

III.4.5 Social aspects

Another aspect of the demand for new transportation systems is the service that could be provided for disadvantaged sections of the population who may not have a car available and who may suffer from limitations in their ability to use some forms of public transport (e.g. the old, the young, physically handicapped, etc.). Such social questions may affect patronage to some extent but would also have an important bearing on the relative evaluation of possible new systems. In studies where social implications are important it may well be necessary to stratify the travelling population in considerable detail in order to identify the sections deriving benefit. This reduces the possibility of simplification.

III.4.6 Investment decisions

The problems that have been outlined arise from the attempt to utilise conventional assessment methodologies for the strategic assessment of new modes of transport. Difficulty in this area derives from the different nature of the questions regarding existing modes. For a new mode, money would be needed for the development of the relevant technology. For existing modes, money can be invested directly in building and operating specific networks in specific places. Much of the information required in each case is the same (i.e., the operational and economic characteristics of the system if it were built), but the assessment of investment in new modes requires a degree of generality that is not necessary with existing modes. Assessment of the total potential market for a new system is needed for the decision on whether or not to invest in developing the system. The implication of this requirement is the need to be able to assess the potential transport systems in a wide variety of urban situations. This is, again, an area in which relatively little work has been done; the detailed nature of the normal land-use/transportation study has not provided generalised relationships relating travel demand to urban parameters. Such work is needed to permit an assessment programme for new modes of transport. The detailed studies which have been carried out can be used to assess the degree to which actual urban situations are amenable to generalisation. If this type of approach is successful, the generalised relationships could then form the basis of testing the performance of alternative transport systems. From this information estimates of the potential market for the various systems can be made. This would be the first round in an iterative procedure to account for the effects of economies of scale.

III.4.7 Conclusions and implications for simplified models

It has been shown that conventional four-step procedures are not well adapted to the study of new modes which might have an important influence on travel. Equilibrium models are better in this respect but more work is needed on the descriptors of user behaviour for satisfactory modelling of the effect of new systems on modal split. The shortage of data on non-vehicular traffic is a problem when studying short distance transport systems. The need for assessments of new modes to cover the whole market potential and to include research and development costs make these studies different from those for existing modes and require model results in a range of possible circumstances. For this reason the strategic type of model is particularly important for new mode studies but it is apparent that such models should retain considerable detail with regard to mode characteristics. CRISTAL is an example of a model in this class but is limited in that it applies only to a large area of London. Development of network simplification models which are usable in a range of urban situations would be desirable.

III.5 Traffic Restraint

III.5.1 Model requirements

The study of restraint of traffic either by means of special charges or by physical limitations make particular demands on the models employed. Features which make a traffic model suitable for road traffic restraint studies include:

- a mechanism for re-routing and choice of mode,
- a mechanism for trip generation and suppression,
- sufficient detail as output to allow different types of restraint to be analysed and evaluated on a common basis,
- an adequate description of peak loading and the effect of peak spreading.

The essential mechanisms that should be provided include:

i) provision for differential user responses to charge levels,

ii) a mechanism to ensure that trip demand changes are consistent with traffic changes.

Desirable features form an extensive list, but some are of much greater importance and relevance than others; i.e.,

- response of modes of transport to restraint measures affecting them only indirectly;

- the identification of different types of trip and traveller to allow benefit and disbenefit transfers to be studied;

- a specific mechanism to show the effect of physical constraint on trips.

III.5.2 Short and long term effects

Traffic restraint measures have both short and long term effects and the requirements for study differ markedly between these two time scales. Short-term responses to traffic restraint measures specifically require a network wide model responsive to "small" and local disturbances at the traffic management scale. Long term responses to traffic restraint require a substantial understanding of the location criterion for activity systems in the area affected directly or indirectly by the restraint measures. It is therefore appropriate to distinguish between long and short term response models by the depth and complexity of the travel generation and distribution mechanism used.

Short term models do not require sophisticated sub-systems concerning activity location and can therefore be greatly simplified in this part: however, the ability to respond to very small local changes with some fair measures of geographical representation is critical to short term problems, and no simplification is admissible that reduces the effectiveness of the geographically-specific balance between supply, demand, and travel responses. Long term models require comparatively little geographically specific network details, and can be drastically simplified in their detailed representation: however the supply, demand, travel response mechanism is still of central importance, and the activity shifting and altered attractiveness and accessibility sub-models brook no simplification. The underlying distinction between long and short term models that allows us to simplify in specific areas is the length of time taken for location of activity to respond substantially to the altered conditions. For a brief time span the redistribution of flows and overall reactions on travel demand are central issues: for the longer term the alterations induced in activity patterns take over.

III.5.3 Simplified models

An example of a suitably simplified model of short term response is the RRLTAP restraint model(18,30,31,91,92). This adopts a simplified travel demand response sub-model, but makes stringent demands on the network and travel behaviour response sub-models to achieve a high degree of consistent sensitivity and detail in the simulation of locally applied charging policies. The model is described in the references given above and is distinguished from others in the close attention paid to equilibrium balancing of road against road, route against route, and trip costs against trip demands. Restraint models are designed to compare fiscal policies such as parking charges, road pricing in local areas and under different pricing restrictions, cordon charges, and mixtures of all these pricing strategies. As the actual level of charge in each small region of a network under the prevailing limitations on area and type of charge is of considerable importance, a special model for estimating the internal reliability of each network benefit figure obtained was needed. With known error bars on each point the reliability of the conclusions drawn from a systematic study of local and overall fiscal restraint policies may be monitored. Other models concerned with a fairly short term viewpoint are the West Midlands Transportation Study (where parking charges were studied as a means of adjusting the overall level of travel demand), and the use of the SELNEC(53) transportation model to study the effects of traffic restraint. Both WMTS and SELNEC were also concerned with the longer term problems, and the SELNEC model in particular has a structure suitable for both time scales. However, these models cannot be regarded as "simplified", as the full apparatus appropriate to a detailed transport planning system was included in both cases. The London Transportation Study introduced a simple and pragmatic model of longer-term traffic restraint; this was carried out by a linear programming procedure and while the results were helpful and indicative, the technique as applied was at that time insufficiently developed to provide a firm basis for forecasting. Of these models the SELNEC one has proved to be that best adapted to a "simplified model" mode of operation.

In the United Kingdom some smaller models of the SELNEC type(53) have been set up and indeed applied to the SELNEC area, e.g., SIAM, IMPACT, COMPACT. All these are basically similar, but IMPACT has been used extensively for experiments with different modelling methods and therefore this name covers a considerable number of different programmes. All these models are of basically the SELNEC type, and have been simplified by limiting the number of different modes and trip purposes rather than by reducing the size of area considered. The size has been cut down quite severely in COMPACT, but this is no longer typical of this family of models.

One line of approach to the overall strategic problem is the CRISTAL model (see II.4). The prime feature of long term relevance is the demand model used to characterise trip generation and distribution in a manner wholly consistent with the evaluative framework.

This compromise in representation allows the model to be used to study multi-modal policies over the system as one integrated entity, and therefore provides a half-way house between short term and long term models.

No location shift effects are built into the CRISTAL demand model, and the differential influences on location and consequent demand patterns require very much better understanding. The work described in section III.2 demonstrates the importance now being attached to these aspects.

Another model which is appropriate for use in analysis of restraint policies is DODOTRANS. Because of its explicit equilibrium structure it is particularly well suited to reflect the issues outlined earlier(29).

III.5.4 Some results

Typical results of a comparative study of restraint policies over a network are shown in figure 8 on two axes:

- net benefits to the community,
- the revenue changes induced by the specific proposal that generates this benefit.

Desirable policies would produce large benefits at the expense of small revenue generation (thereby reducing the costs and social implications of returning such revenue as a transfer payment). Significant features of this illustration are:

a) the importance of the error bar measures for each point;

b) the relative robustness of comprehensive parking policies to price level errors;

c) the consistent effectiveness of road pricing schemes in achieving significant benefits with minimal exaction of revenues levied on the users;

d) the notable effectiveness of joint road building and road pricing policies;

e) that special licensing (or cordon pricing policies) is critically dependent on the precise geographical location and size of cordon, but can become a useful and effective policy.

While this class of model is of considerable value in short term response assessment, severe difficulties lie ahead in the long term response models where the network

Figure 8

COMPARATIVE EFFECTIVENESS
OF A RANGE OF RESTRAINT POLICIES

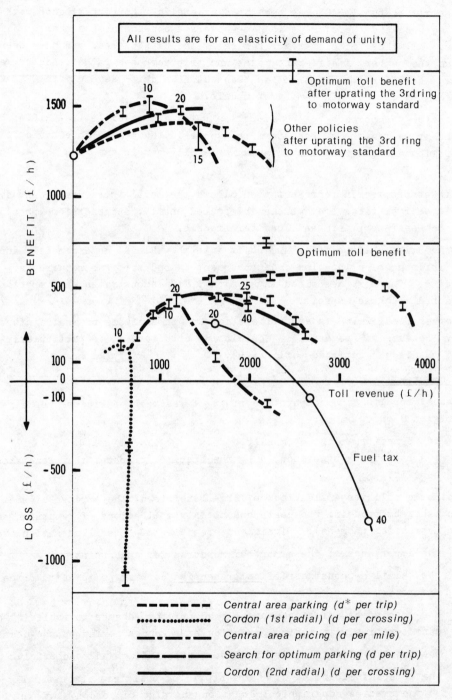

Points marked are in units shown, except that fuel tax is in multiples of present level.

* 2,4 d = 1 new penny.

refinements referred to in III.5.3 are of little value, and a significant improvement in modelling and understanding must be obtained at the expense of detailed representation of local network.

III.5.5 <u>Conclusion regarding simplification</u>

While the simplifications laid out in III.5.2 are likely to prove adequate for fiscal restraint policies, physical restriction schemes require more detailed models of the linkage of part-journeys into round tours, as the altered balance of opportunities for activity is not adequately represented in current models. A possible simplification that might be introduced is the complete removal of geographical representation in favour of more detailed analysis of conditional linkages between trips, modes, purposes and activities.

Clearly, both long and short term responses are of importance, and one must be able to analyse both fiscal and physical restraints separately and in combination: if these issues are separated on the lines suggested above, one may use drastically simplified models that may permit the analysis of each situation in an appropriate manner and isolate the vital issues and criteria.

The essential feature of models exployed for the study of traffic restraint should be the incorporation of an explicit demand model in an equilibrium approach which can allow generation, distribution, modal split and path choice to be all effected simultaneously.

The above issues and simplifications are all essentially cross-sectional or equilibrium view points. Severely simplified models responsive to the interaction through time of location, mode usage, and travel response may well provide the insights necessary to achieve the objectives determined for a long term response to traffic restraint.

III.6 <u>Parking</u>

Traffic and parking are closely connected. A private car is normally in motion only 1/24th of the day. Every trip ends with the vehicle being parked for a long or a short period of time, depending on the purpose of the visit at the trip end. Parking may thus be classified by the duration and by the activity forming the reason for parking. A general forecast model describing both traffic and parking can theoretically contain total trip chains of the type stop-move-stop-move-stop, giving the position and time for the changes in state of motion. When the chains are summed up, the total number of stops and moves with time durations would then be obtained together with time variations. Modelling of this type has however not been done as it would be very complicated. Furthermore the uncertainties in the less stable elements (for example, shopping and recreation trips which might have no trip end with parking) would also increase the risk of unreliable forecast results.

III.6.1 <u>Parking characteristics</u>

A brief description of how parking conditions influence the traffic situation and why they have to be included in the forecasting process is necessary. Parking conditions and characteristics will however vary extensively depending on a number of factors, such as the geographical and historical setting, the degree of economic development and motorisation, the social and judicial structure as well as parking policy and quality of public transport systems. All these are different from country to country and from town to town.

With increasing economic growth and mobility, pressure is experienced at focuses of attraction e.g. centres of shopping and employment. As the central area of a city is the point of focus of public transport and private cars, shops and offices expand at the expense of dwellings and a CBD is developed. Although in recent years no substantial increase in total floor area in some city centres has been experienced, office floor area has continued to increase, while other activities decrease(54). To cope with an increasing share of individual transport and parking difficulties, planning to handle present and future parking demand is necessary.

A common first step in regulating parking is to limit the duration and prohibit parking especially in environmentally sensitive areas. A further step is the introduction of charges. Next, the arrangement of special parking facilities with increased fees is needed to cover costs and facilitate a separation of allowed parking duration. This implies that a series of legislative measures is implemented to cope with illegal parking.

The average city will today be characterised by zones with different parking conditions. These occur automatically as a result of the interaction of demand and supply and may also be controlled by local traffic regulations.

The following table illustrates data from Stockholm.

Table 4

Parking Data from Stockholm

Stockholm 1972(56,57)	Region	Inner Town	CBD
Inhabitants	1 300 000	230 000	6 000
Households	530 000	130 000	4 500
Employed	643 000	295 000	90 000
Households with car	55 %	30 %	
Parking places, day, total		60-65 000	8 500
On streets, with charge		25 %	15 %
On streets, without charge		7 %	-
Parking garages		40 %	80 %
Lots, yards, etc.		28 %	5 %
Place for parking, all trips			
On street	45 %	64 %	
Parking garage	4 %	9 %	
Lots, yards, etc.	46 %	27 %	
Paid charge when parking			
None	94 %	86 %	76 %
0 - 2 S.Kr.	5 %	12 %	16 %
> 2 S.Kr.	1 %	2 %	8 %

Little is documented about the effects of charging policies. It is however becoming evident that one of the most important means of managing traffic in central areas will be the parking policy. Based upon 10,000 interviews held in 1968 in Stockholm it was possible to obtain data for simulating the effects of various strategic measures in traffic. A doubling of the parking charge in the central area (no free parking) would reduce the proportion of commuters going by private car from 32.4 per cent to 30.8 per cent. This may be compared with the effects of completely eliminating fares on the entire public transport system which would reduce the proportion to 29.4 per cent. These calculations have later been verified by actual practice after the implementation in 1972 of doubled parking charges and the introduction of the so-called 50-card. The latter means that for a monthly fee of S.Kr.50 unlimited free travel is allowed on the public transport system. At present the card is used by 75 per cent of the commuters to the central area.

III.6.2 Parking and four-step model

In conventional modelling using the four-step model the concept of parking has been dealt with in a rather unsatisfactory way. Parking conditions may have been included in the distribution step by including time for the search for parking space, the parking operation and the pedestrian time as well as the cost for the parking which may be varied from zone to zone (when using generalised cost as a distribution tool). Also in the modal split phase a similar procedure has been adopted. In detailed studies the actual location of various types of parking facilities may have influenced the road assignment step as well.

Having the number of car trip ends at each zone and making assumptions about parking duration, based on curbside studies of parking, the "demand" for parking spaces in an area may be calculated. More often, however, the amount of parking has been determined by directly using the data about land use of the area, distinguishing between dwellings and areas with frequently patronised shopping. Based on studies of current behaviour through parking questionnaires, relationships have been established between land use, frequency of car visits, parking duration and level of motorisation. It has been possible to demonstrate a relationship between size of town, peak hour traffic and maximum number of cars parked in the central area as shown in figure 9.

Figure 9

RELATION BETWEEN SIZE OF TOWN, PEAK HOUR TRAFFIC,
AND MAXIMUM NUMBER OF CARS PARKED IN THE CENTRAL AREA

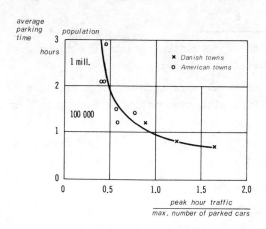

As parking facilities have not satisfied environmental requirements and traffic congestion factors have become more evident, the inadequacy of the handling of parking in planning has become apparent. Parking as a restraint factor must, as the gap between demand and supply increases, influence all the stages in the four-step model. Some attempts to study this might be done by feed-back processes, a possibility which can be developed further. There have been cases where the excess demand for parking in a central zone has in the planning process been transformed into an increased number of public transport trips equivalent to the unsatisfied car trip ends. Even if an iteration process is developed to handle parking in a better way, some serious limitation will still remain. Thus, combined trips, park and ride, parking duration control considering trip purposes, charging policies and enforcement efficiency are difficult to handle adequately in the conventional modelling.

III.6.3 Parking and demand models

Parking problems may more easily be included in demand/supply models. By introducing the marginal cost for each additional parking space as well as the actual elasticity of demand, an equilibrium is obtained in one process. Effects of parking charge policies and travel response to various cost levels for the parking arrangements may easily be evaluated as part of strategic planning.

The need for surveys to supplement existing data is apparent. Some of the short-comings of the conventional models as regard to parking will certainly also apply. Examples of such difficulties are the effects of car pooling arrangements, enforcement and environmental effects.

III.6.4 Possibilities for simplification

The degree of simplification in models for studying parking problems will have to correspond to the requirements of the user. In turn, these requirements should be related to the planning stage as will be suggested in the following sub-chapter.

III.6.5 Planning for traffic and parking

Considering the demands of long term strategic planning, short term planning as well at present operating policies, the following simplified procedure is suggested.

Level 1. Strategic planning

-   Total number of trips to central areas as a function of predicted land use and of local residential structure;

-   Expected amount of total parking area as a function of land use with special regard to activities with high visiting frequency (commercial and retail business);

-   Cost level for providing parking facilities as well as funds available for traffic and parking.

Level 2. Tactical planning

-   Numerical evaluation of expected number of car trips to CBD zones;

-   Numerical relationships between car traffic and parking spaces as a function of zone characteristics;

-   Evaluation of parking space availability considering parking arrangement costs, environmental limitations and desirable street system capacity;

-   Policy decision concerning the degree of parking supply relative to the demand;

-   Possibilities for alternatives to private vehicle traffic to the restricted areas such as increased public transport capacity, park and ride arrangements, other modes, etc.;

-   Suitable distribution of parking facilities between expensive central areas and less expensive peripheral areas (subcentres, shopping centres).

Level 3. Operation

-   Parking charging policy to be related to the desired level of car usage for the area;

-   Parking duration policy which must be a function of the characteristics of the area (brief shopping visits compared to employees 8-10 hour stays);

-   Local traffic legislation to enforce the desired parking policies;

-   Judicial needs to cope with violation (roadside parking, parking ticket obligation, car removal possibilities);

-   Before and after studies to establish relationships between parking policy and resulting effects as regard to duration, charging, enforcement and other modal possibilities.

Strategic planning mainly concerns control of land use. The parking facilities must at this level influence the future intensity of car traffic generation so that parking arrangements can be economically feasible. This implies that simplified techniques may be used in forecasting.

Parking will be dealt with in more detail in tactical planning. Parking policies do at present tend to be one of the tools to keep car traffic low in environmentally sensitive areas. A balance must at this stage be obtained between forecast traffic volumes, costs for parking and the effects of restrictions on the desired function of the area. Account must also be taken of the fact that parking restrictions only reduce trips which terminate in the area and may in fact encourage through traffic.

Data concerning the effects of various parking arrangements and enforcement are limited. Only data concerning the desired amount of parking related to land use seem to have been documented. The requirements tend to be concerned with the effects of various measures and the secondary effects on land use when restrictions are imposed. Such data collection is necessary for both conventional and demand models.

III.7 Pedestrian Aspects

Predictions of pedestrian trips are needed by transportation planners for use in the planning of pedestrian ways, precincts and interchanges and for two purposes connected with the prediction of vehicular trips. These are the estimation of future diversions between the pedestrian and vehicular mode and the calculation of the walkers' contribution to trip ends.

Walking tends to attract much less capital investment than vehicular modes of travel; therefore, simple approximations have sufficed for many planning purposes.

It is of interest to note that the detailed modelling within an area of the generation, distribution and assignment of pedestrian trips is formally much the same as the more common calculations for vehicular modes and that the cost and effort are about equivalent to the vehicular ones for a small town(58). Thus, the advantages of simplification here are great.

Approximations may be divided into three classes. The first kind are made by coarsening an existing representation of a system, for example, by dividing an area into a few large zones and applying procedures commonly used with many zones. The second kind are obtained by taking the exact formulation of a system's behaviour and then making an approximation too gross for a detailed calculation but adequate for a rough one. Thus, Edwards and Shipley(59) replace the time distance between zones by the crowfly distance multiplied by a constant route factor. The third way is to estimate the quantity of interest by a more or less empirical function which does not represent the logical consequence of travel behaviour in the district but is a simple combination of factors which common sense suggests might influence, or at least be correlated with, that desired quantity. The linear function of cell attributes by which Percivall and Sandahl(60) predict the number of people in each small cell of a town centre are an example of this method.

All three types of approximation have been used in the modelling of walk trips but the arbitrary or empirical function seems to have been the universal favourite.

A selection of examples of these methods is shown in Appendix 5.

IV

ACCURACY CONSIDERATIONS

IV.1   Introduction

A key question which arises in a discussion of ways of simplifying models is that of accuracy. The objective of this chapter is to discuss the question of errors and accuracy.

Two major sources of error must be considered in every transport analysis. First, there are always errors in the travel forecasts (or for that matter in the results of any models), due to a variety of factors such as sampling errors, estimation (calibration) errors, differences between future assumptions used in the forecasts and actual events, and the essential randomness in human behaviour. Second, there are "errors of analysis": which transportation policies are studied, and which are not; the degree of detail in the alternatives studied and the effects; the number of alternatives and variations of alternatives which are studied, etc. Therefore, the transport analyst must explicitly consider the way he allocates his study resources to both types of errors, as there is an important trade-off between the two. In one extreme, a very detailed and precise model can be constructed, but with so few study resources left that only one or two alternatives can be studied. At the other extreme, a highly simplified model could be constructed, to allow analysis of a large number of alternative transport strategies, but with little resources devoted to calibration of the model. Between the two extremes lie many variations. In any consideration of accuracy, the analyst must be aware of these trade-offs, especially in the development of simplified models.

Since there will always be some errors in the travel forecasts, the analyst should do studies of the sensitivity of his results to alternative assumptions about model parameters and even model structure; resources must be allocated to these studies. Also, explicit measures of the range of uncertainty in each forecast should be established and reported in all analysis results.

In the following sections, the range of magnitude and sources of errors in the conventional models are discussed. In developing simplified models, the analyst should keep in mind the relative errors in the conventional four-step models and the trade-offs discussed above.

IV.2   Sensitivity Tests

A model is merely a representation of reality; it is always affected by incidental or systematic errors. It is necessary to minimise these errors in order to adapt the model as closely to reality as possible. This adaptation often progresses iteratively in several steps or cycles.

Once a concept for the model has been established the determinants of the model have to be defined by an error analysis and a sensitivity analysis in such a way that the total of errors when applying the model will be as small as possible. In Figure 10 the process outlined above is schematically represented.

Figure 10

ERROR AND SENSITIVITY ANALYSIS

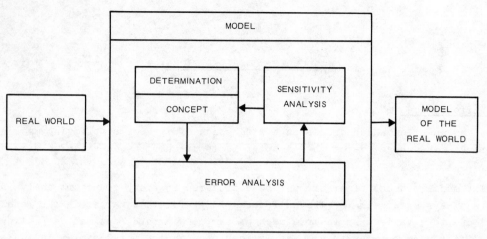

Special consideration should be given to the sensitivity analysis. By this term the following will be understood: if the parameter values change the results obtained when applying the model will change too, while the other influencing factors remain unchanged. Together with a change of parameters the quality of coherence will, as a rule, change as well. The sensitivity of the model to parameter changes which are indispensable for prognoses is clarified by means of a sensitivity analysis. If the results of model calculation change slightly or not at all, it may be assumed that the sensitivity is low, which would be desirable.

The following findings of research carried out on gravity type traffic distribution models may serve as examples for such a sensitivity analysis. In Figures 11 and 12 the sensitivity of these models in consequence to the change of the exponent β is represented(70) on the basis of the resistance function.

$$f(w_{ij}) = w_{ij}^{-\beta}$$

vid. Figures 11a, b and c where the sensitivity of the model is described by an empirical coefficient of correlation X.

In the study(71) the exponent β in the resistance function

$$f(w_{ij}) = w_{ij}^{-\beta} = w_{ij}\xi \ G\lambda$$

has been assumed to be a function of journey time and accessibility. A characteristic representation of the model sensitivity consequent to a change of the journey time exponent ξ is shown in Figure 12.

The continuity of the plotted curves and their position is, for instance, dependent on:

- model type
- size of city
- size of traffic zones and their function
- purpose of journey
- time of day.

From the foregoing studies(70,71) it was concluded that formal error analyses only (e.g. according to standard deviation) is not sufficient to estimate the reliability of a certain model and sensitivity tests are vital if conclusions regarding the validity of model results are to be drawn.

Figure 11

DEPENDENCE OF THE EMPIRICAL COEFFICIENT
OF CORRELATION χ ON β

$$f(W_{ij}) = W_{ij}^{-\beta}$$

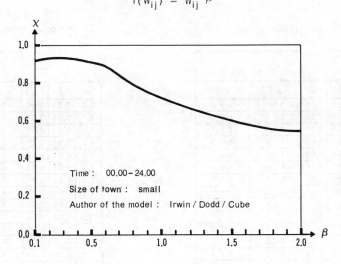

Time :   00.00 – 24.00

Size of town :   small

Author of the model :   Irwin / Dodd / Cube

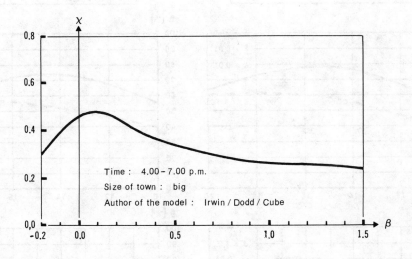

Time :   4.00 – 7.00 p.m.

Size of town :   big

Author of the model :   Irwin / Dodd / Cube

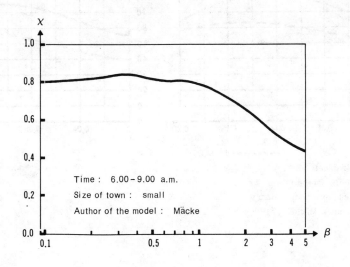

Time :   6.00 – 9.00 a.m.

Size of town :   small

Author of the model :   Mäcke

Figure 12
COMPARISON OF CALCULATED AND SCORED VALUES F_{ij}

$$f(W_{ij}) = W_{ij}^{-\alpha} \, W_{ij} \, \xi \, G_i \, \lambda$$

$6^{00} - 10^{00}$ a.m.

α	ξ	λ	B [%]	
			i = 60	i = 65
1,0	0,01	0,01	53,1	43,4
1,2	0,01	0,01	67,0	46,3
1,4	0,01	0,01	76,7	55,8
1,6	0,01	0,01	77,8	60,0
1,8	0,01	0,01	72,9	61,6
2,0	0,01	0,01	62,5	60,8
1,0	0,02	0,01	56,5	45,5
1,0	0,04	0,01	63,3	—
1,0	0,06	0,01	69,6	53,6
1,0	0,08	0,01	75,1	57,1
1,0	0,10	0,01	78,9	60,0
1,0	0,12	0,01	80,8	62,0
1,0	0,14	0,01	80,2	63,0
1,0	0,16	0,01	—	62,8
1,0	0,20	0,01	63,5	58,3

$3^{00} - 7^{00}$ p.m.

α	ξ	λ	B [%]	
			i = 60	I = 65
1,0	0,01	0,01	55,1	39,1
1,2	0,01	0,01	67,5	46,3
1,4	0,01	0,01	74,0	52,1
1,6	0,01	0,01	71,8	56,0
1,8	0,01	0,01	60,6	57,7
2,0	0,01	0,01	41,9	56,9
1,0	0,02	0,01	58,3	41,2
1,0	0,04	0,01	64,5	—
1,0	0,06	0,01	69,8	—
1,0	0,08	0,01	73,8	—
1,0	0,10	0,01	75,7	55,9
1,0	0,12	0,01	75,2	58,0
1,0	0,14	0,01	72,0	58,9
1,0	0,18	0,01	—	57,0
1,0	0,20	0,01	47,0	—

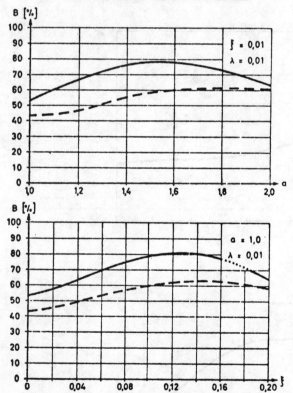
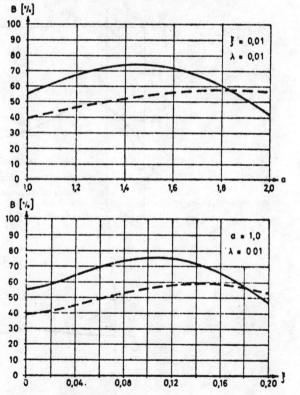

—— i = 60   - - - - i = 65

i = No. of traffic cell.
B = Coefficient of determination
Individual transport in Ruhr region.
Author of the model : Mäcke. (71).

IV.3 Errors in Model Components

The following sections give some examples of the principal errors in traffic model components. A definition of terms and further examples of errors are given in Appendix 6.

IV.3.1 Errors affecting estimates of traffic generated by households

Errors due to different sample size

An example of the way sample size affects the accuracy of calibration data is provided by results from London (3 million households, divided into 108 categories). For traffic generated by a category of households classified as "important", e.g. "2 persons, 1 car, medium income" (representing 300,000 London households), the error was as follows:

Number of households interviewed	600	3,000	18,000
Per cent of households interviewed	0.2	1.0	6.0
Error in traffic generation estimate (95 per cent limit)	± 4%	± 2%	± 0.75%

Errors due to the inaccuracy of the model

It is possible to devise models and adjust them in calibration to give almost any degree of matching to measured data. However, as a practical example in the Glamorgan study(83), the mean numbers of observed and calculated trips per household (for households with one car) compared as follows:

Employed persons per household	Observed	Model	Percentage deviation of the model from the observed values
0	3.2	2.0	- 36%
1	3.5	3.4	0%
2	3.7	4.9	+ 33%

IV.3.2 Errors affecting traffic from a zone

Errors due to sample size

"Public Roads" of December 1960(84) states the following errors due to the sample size in the number of person trips by car per day from a given zone:

Total traffic volume. Person trips by car per day	1% sample		5% sample	
	Observed error*	Theoretical error**	Observed error	Theoretical error
100	171%	100%	76%	44%
1,000	56%	31%	25%	14%
10,000	18%	10%	8%	4%
100,000	6%	3%	2.6%	1.4%

*   This is the per cent. RMS error.

** This is the theoretical per cent. standard deviation error.

The observed error is thus about twice as great as the theoretically calculated error. The reason seems to be that some of the interviewed persons report too few trips and others too many.

IV.3.3 Discrepancies between measured and estimated screen line traffic

The discrepancies seem to lie between 5 and 9 per cent for medium size towns but may be as high as 34 per cent for small towns. They may also be high where the trips are divided into a number of categories (see Appendix 6).

IV.3.4 Errors affecting the calculation of peak hour traffic

As indicated in Section II.3.6 the effect of approximations in the estimation of peak hour traffic levels can be dominant in traffic studies; in particular the spreading of peak traffic with time is handled by empirical methods which are not necessarily consistent with the models used and should be treated with caution.

IV.3.5 Errors affecting forecasts

Growth factor methods have been widely used but suffer from the need to assume stability in the model and the fact that the zoning process can have a profound effect upon the results. An example of one check of forecast against growth achieved is given in Appendix 6 Section 6. and brings out the large errors in zone to zone trips which are greatly dependent upon traffic volume. The contribution of errors in population and car ownership forecasts and the errors due to the model cannot readily be separated in this example.

IV.3.6 Conclusion

Some data derived from the surveys discussed in this section and in Appendix 6 are plotted in Figure 13.

It will be seen that the error percentage curves for the observations in the following four cases very nearly coincide:

1) London. 6 per cent sample(82).
2) "Public Roads." 5 per cent sample(84).
3) Sioux Falls. Model, calculation of screen line traffic in the analysis year(85).
4) Washington. Growth factor forecast 1955(86).

It should be remembered, however, that the Washington forecast is calculated on the strength of growth factors actually experienced.

The curves for the 0.2 per cent London sample survey and for the 0.25 per cent sample survey of the journey to work in Copenhagen also nearly coincide. With such a small sample, however, the error will of course be very great. In Copenhagen, a more comprehensive survey was in fact carried out later.

The error percentages found by Zuberbühler(87) are somewhat better than in the 10 per cent sample quoted in "Public Roads."

The question of the matching between calibrated model results and measured data is determined largely by the refinement of the process used. This is chosen according to the application of the model concerned. For strategic purposes a broad agreement between model and survey is usually adequate, and this is one of the main ways in which strategic models achieve simplification.

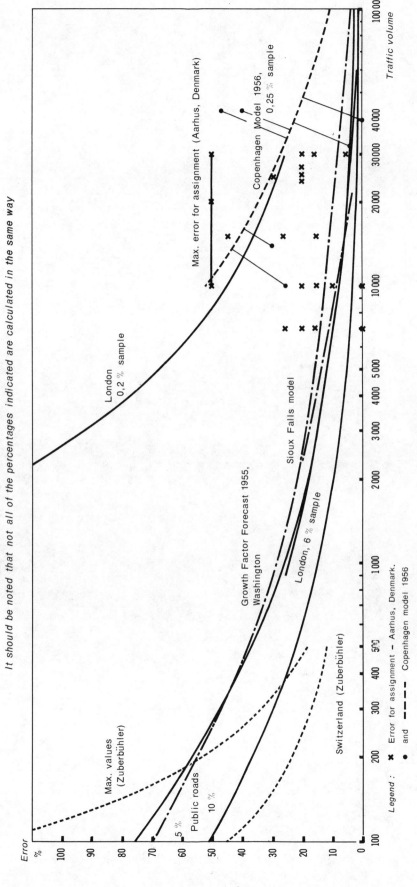

Figure 13

ERRORS AS A FUNCTION OF TRAFFIC VOLUME AND SAMPLE

It should be noted that not all of the percentages indicated are calculated in the same way

V

DATA REQUIREMENTS, AVAILABILITY AND SIMPLIFICATION

V.1   Current Practice: Data Requirements and Availability

V.1.1 Introduction

Whatever the type of model to be used, those responsible for transport studies are generally faced with two constraints:

- budgetary constraints due to the fact that the collection of data necessary for the application of the models constitutes one of the most costly elements of the study;

- constraints as concerns the availability of the necessary information, which can to a certain extent determine the quality of the models to be used.

It is clearly worthwhile, therefore, to devote a special section to the problem of data needed for application of models, with particular emphasis on the practical aspects involved in transport studies, i.e.:

- What data are needed?
- What type of data?
- How and from what source should they be collected?
- What are the resources needed to collect the necessary basic data?

V.1.2 Data requirements

In very general terms, data are used at three stages:

- in the development of the models;
- in the calibration of existing models;
- in the application of existing and calibrated models for forecasting purposes.

The first two stages can, however, be combined: if the transport planner does not know a priori which variables will be taken into account by the model, the experience acquired and the compilation of literature already published on the question will nevertheless enable him to define to a certain extent the kind of information to be assembled. The main problem therefore is the choice of methods of data collection. The technique most often used is random sampling, which is well known and calls for no special explanations here. One might, however, point out that although this technique is relatively simple, its application can prove very cumbersome in practice as it is always possible that the sample may be biased.

As concerns the application of the model(s) for forecasting purposes, the main problem is the projection of the necessary parameters to the target horizon. Forecasting techniques may either be borrowed from other disciplines - demography for example - or they may be specific to traffic studies - forecasting of journey patterns, etc. Other difficulties arise from the fact that the values of certain important parameters are not basic data, but figures resulting from the utilisation of transport networks; these can be particularly hard to estimate.

V.1.3 The type of data necessary

The data necessary may be classified in three groups:

- socio-economic data relating to the area studied;
- data relating to transport networks;
- data relating to journey patterns.

Socio-economic data relating to the area studied

These are, primarily:

- total and working population and students;
- composition of households;
- incomes of households;
- number of motor vehicles per head of population;
- employment, with breakdown as appropriate by category - workers or employees - and/or by type of activity - industrial, administrative, commercial;
- school establishments.

These data are necessary not only for calibration, if not development, of models of the present situation, but also for application for forecasting purposes. They must be available for each small subdivision of the area studied.

Data relating to transport networks

These are essentially:

- length of routes;
- journey time over specific routes;
- capacity of road links;
- cost of journey along these road-links;
- parking capacity.

This information is required both for calibration of models applied in a real situation and for the use of these models for forecasting purposes.

Data relating to travel behaviour

These data are in fact of two kinds:

- factual data relating to travel under the prevailing conditions;
- parameters representing the probable behaviour of the individuals concerned in the future.

Factual data relating to travel under prevailing conditions: these data are, as a general rule, the following:

- number of journeys per day per person or per household;
- trip purposes at origin and destination;
- geographical origin and destination of journey;
- time of departure and journey time;
- mode(s) of transport used.

Parameters representing behaviour under future conditions. These parameters are those included in the mathematical relationships constituting the models used, relating the characteristics of travel as indicated above to the socio-economic data of the area in question and the characteristics of the transport networks serving it.*

V.1.4 Methods of data collection

Data relating to the current situation

Home interviews would at first sight appear to be the best method by which the most varied collection of data may be assembled, both socio-economic data relating to the area studied and factual data concerning journeys made in this area by residents of the zone in which the survey is carried out.

A certain number of comments can however be made concerning household surveys (interviews):

- in the first place, if the survey extends far out from the centre of the conurbation being studied, the frequency of daily journeys per person in the zone studied tends to be substantially reduced as a result of the increase in distance, although in absolute terms the number of such journeys in the zone covered by the study need be by no means negligible; in these circumstances the results of household interviews may be poor, so that it is preferable to take two separate residence areas and adopt different methods of data collection with respect to the residents of these two areas;

  - the first area comprising the zone being studied and its immediate neighbourhood, where the frequency of daily journeys by residents of the zone studied is high, and within which the household interviews will be carried out;

  - the second comprising the remaining area of the total region studied, where residents will no longer be interviewed at home, but when crossing the boundary of the area where the household survey is carried out; the best way to conduct such a "cordon survey" would appear to be to question users leaving the conurbation, at such exit points as stations for rail travellers, for example. In order to determine the border lines of these two areas it might be necessary to carry out a pilot study along one or more radial axes of the conurbation being studied;

- in the second place, the high cost of surveys by households leads one to look for other methods of assembling certain types of information. It is, for example, to be noted in this connection that information collected at the time of the general population censuses carried out periodically in each country includes data relating both to place of residence and place of work, to the degree of motorisation, to the type of employment and to the conditions in which journeys of the homework type were made (mode of transport, journey time, etc. . . . .). As this information is generally analysed with different objectives to those of a transport study, it then becomes necessary to conduct a survey of households or individuals on the census register in order to reconstitute the necessary information at the appropriate geographic level.

---

* For example, the parameter λ in the gravity model

$$n_{ij} = K \frac{a_i \, e_j}{t_{ij}^{\lambda}}$$

which associates the number of home/work trips n_{ij} between zones i and j with the number of working residents a_i of zone i and the number of employments e_j in zone j, and with the trip time t_{ij} from zones i to j.

Special surveys may also be carried out in order to collect certain types of information, for example, data on certain special trips (i.e., trucks and taxi trips and public transport using public roads). Similarly special surveys may be conducted regarding parking, possibly distinguishing between:

- on-street or near street parking;
- off-street parking,
  - provided by firms for their personnel, visitors or clients,
  - provided and rented by special public or private parking companies.

The following are some examples from existing studies:

(a) <u>Number of Trips Per Car Per Day as an Average for the Whole City</u>

U.S.A.: Reference 23, based on studies in 50 American cities, reports trip rates which varied between 2.0 and 7.2 trips per car per day in a way that appeared unrelated to city size or car density.

Great Britain: Some results show a variation of from 3.9 to 6.2 trips per car per day.

Denmark: In four cities, the rates were found to vary between 3.4 and 4.2 trips per car per day.

(b) <u>Number of Trips Per Car Per Day for the Different Zones in a City</u>

An investigation carried out in the Swedish city Gävle showed that the number of trips in the different zones varied from 1.4 to 4.2 trips generated per car per day. There was no relation between the trip rates and the distance of the zone from the city centre.

For the Danish city Kolding, the following relation was found between trip rate and zone location:

Zone location	Trips per car per day
City Centre	3.6
Suburbs	4.4
Rural	3.3

(c) <u>Number of Trips Per Car Per Day for Different Household Types</u>

The number of trips per car per household classified according to number of of cars and income from London data is shown in figure 14. This shows mean values, although there is quite a scatter about these means.

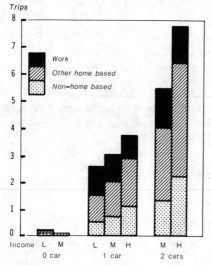

Figure 14

NUMBER OF TRIPS PER CAR PER DAY
FOR DIFFERENT HOUSEHOLD TYPES

Data relating to the future situation

Socio-economic data. Socio-economic data are derived from various disciplines which are not specific to the field of traffic study proper. Traffic forecasts are required not only for the total conurbation being studied but also for each of the various zones adopted for the purpose of the study.

Data available are generally those from urban development plans which are often limited to reserving available space both for housing and for the different sectors of activity, whether existing or planned.

With regard to demographic data, the method most often used consists of two stages:

1. to determine the general evolution of the total area being studied, or of the region of which it forms a part, using a classical forecast method such as following up the population trends by sex or age group;*

2. to make a forecast of individual zones taking into account:

   - the global trends determined elsewhere,
   - the recent trends observed by zone or group of zones,
   - the objectives set in existing urban development plans.

Similarly, employment data are treated in two stages:

- the first consists of determining the global trends and, if necessary, differentiating between activity sectors;

- the second leads to the prediction of employment by zone, taking into account, on one hand, the global trends mentioned above, and on the other, recent evolutions and the relative size of the areas devoted to employment purposes in urban development plans.

Finally, motorisation is predicted on the basis of a logistic trend,** the asymptote of which will be equal to the saturation rate of the population considered.

_____

*   i.e., by simulating the evolution of the age distribution.

** i.e., a function of the form $y = a/\ell + be^{kt}$.

Data relating to future car density in different zones. It is possible to relate vehicle density to "housing density" and "accessibility to public transportation" for the particular zone in question. These variables and relationships are relatively easy to compile and use in a prognosis.

Another possibility would be to include average or mean income and average age of residents in the prognosis as well. These additional variables, however, are difficult to predict for preplanned housing construction.

Data relating to future trips per car per day. It is expected that the number of trips per car per day will decrease with an increase in the percentage of 2 car families. Several investigations undertaken show the value of trips/day to be lower for 2 car families than for single car families.

	Number of trips/car/day		
	Kolding	Copenhagen	L.A.
Households with 1 vehicle	4.1	3.8	4.8
Households with 2 vehicles	3.0	3.1	3.7

The values given above for the three cities mentioned are in reasonably good agreement with each other.

Data relating to the use of transport networks. Existing transport plans can make it possible to determine fairly easily certain elements, notably the length of routes between pairs of origin-and-destination zones. Nevertheless, in order to apply models it is often necessary to determine journey time or cost for the user, in the form of a linear combination of journey times and distances. Although at first sight this type of information seems fairly easy to determine for public transport with its own right-of-way - the underground railway for example - it is much more difficult to obtain such information for modes of transport using the public highway, since journey time on a given road depends not only on the traffic using that road, but also, to a certain extent, on the traffic using other roads in the same network, although one has no information on this traffic and little or no knowledge of traffic flow.

In the case of a strategic transport study it may, however, be possible to find a normative solution to this problem, i.e., to specify a priori a minimum average journey speed on the network, this speed being differentiated if necessary according to categories of roads, on the assumption that journeys are effected on the route which satisfies the criterion of shortest geometric distance between pairs of origin and destination zones. An iterative method can then be used to revise the trip time data for the various routes of the network and the corresponding traffic distribution according to the same origin and destination pairs.

This method is, however, rather unwieldy and an approximation methods might be preferred which consists of correcting the actual trip times whenever either the traffic volume variations or the alterations of the conditions of using the network suggest it.

As concerns parking capacity allocated to private motor vehicles, in so far as these capacities are important for the future situation (and not a result of the study), it is necessary to define the future parking policy beforehand. On the basis of this parking policy being included in the urban development plans in the form of space reserved for parking or of housing density to be respected, it is possible to determine the capacities for long-term parking associated with home/work trips or for medium and short-term parking associated with other trip purposes.

Parameters representing journey behaviour. The values of these parameters will
have been obtained by calibration of models, using data relating to the reference situation,
or the present situation. These values cannot accurately be used for a future situation
unless the results of earlier studies provide the means to assess their stability; if
the hypothesis of stability cannot be confirmed, it will be necessary to forecast trends
and then use the values calculated in this way.

In this connection it should be noted that observed values for a given parameter
may not be available and, even if they are, the conclusions regarding the stability of
parameters may be different according to study areas.

It does not seem possible as yet to arrive at definite answers to this problem
without future studies on the basis of which an adequate method could be established.
Under these circumstances it can only be noted that, in the majority of studies underway
or carried out so far, the principle of the stability of parameters has been taken for
granted.

V.1.5 Resources needed for the collection of basic data

In the present paragraph an attempt is made to estimate the financial efforts
entailed in the collection and preparation of data needed for the study, i.e. for the
following tasks:

- organisation, preparation and control of various surveys;

- costs entailed in carrying out interviews, codifying data and transcribing them
  as necessary (perforated cards or magnetic tapes);

- verification of the existence, plausibility and coherence of the transcribed
  data and correction, if necessary;

- control of the interviews ensuring that the collated data are representative
  of the total population;

after which the data can be processed using the computer programme(s) for analysis.

In order to estimate the resources needed, the Group made an inventory of costs of
data collection in a sample of urban areas of varying sizes. Examination of this information has shown that amongst the various parameters characterising the urban area (total
population, employment, number of zones, etc.), the total population of the area under
study seems to correlate best with the costs involved. The costs of data collection in
Brussels has ranged between 0.5 to 1 FF per inhabitant, but largely depend on survey sample
rate; surveys elsewhere have been considerably more expensive. It would, however, be more
instructuve to provide information on the costs of the three main aspects of data
collection:

- demographic and socio-economic data,
- travel behaviour characteristics of the population;
- network and model calibration requirements.

The information available has not enabled the Group to cost these three aspects. The
costs for each can vary widely; for instance, demographic information is frequently available for national population census; the travel characteristics for strategic modelling
purposes can make use of generalised information and the data required for model calibration is entirely dependent on the detail included in the model.

V.2   Data Simplification in Current Models

V.2.1   Introduction

One way in which simplifications can be brought about in current urban traffic models is through changes in the data collection and analysis stages. These changes can be obtained in the five main ways described in the remainder of this section.

V.2.2   Aggregation

Aggregation means combining several elements of models in order to reduce the number of items to be explicitly analysed.

This provides an apparent increase in accuracy and may well be necessary to eliminate the misleading biases which can occur in highly disaggregated information. However, for strategic studies which are frequently concerned with policy issues, there is a tendency to require information on the effects on particular sectors of the population. Thus, the objective of simplification in order to have flexible strategic models tends to conflict with the requirement for this information and a compromise is necessary to keep data requirements within limits. It is important to be sure that segmentation, if adopted, is relevant to the output required and does not stress the input data unduly since a very efficient direction for simplification involves the suppression of irrelevant segmentations.

Aggregation of market segments

In principle, there are many different ways in which the population of potential travellers can be grouped into "market segments". For example: stratified by income, car ownership per household, family size, etc.; by trip purpose; by time of day (peak, off-peak, etc.); etc. These market segments can be aggregated into a smaller number - e.g., the three trip-purposes classification often used (home-based work, home-based non-work, non-home-based) - or even into one segment. However, since each segment differs in its travel behaviour, aggregation of segments requires care and thought in case too much accuracy is sacrificed.

Aggregation of zones

Aggregation can be accomplished by combining groups of traffic zones into one zone (for each group). The degree of disadvantage in zonal aggregation depends to some extent on the network assignment technique used. If "all-or-nothing" procedures are used, then zonal aggregation can significantly distort the travel pattern. If, however, proportional assignment, incremental assignment, or other multi-path assignment approaches are used, the distortions are likely to be far less significant.

Network aggregation

In this approach, several links (or a sub-network) are represented by one or two "aggregate" links. The extreme case is the use of "spider networks", in which each zone is connected to each adjacent zone by only one link.

Network aggregation is done implicitly all the time - e.g., whenever analysts in coding a network decide to leave out certain streets or alleys.

Various procedures can be proposed for network aggregation, but little is known about their advantages and disadvantages.

It seems likely that selective aggregation will be most promising. For example, in studying the various options for a particular public transport corridor, it may be reasonable to aggregate highly portions of the network elsewhere in the region, while preserving detail in the network close to the corridor of interest.

V.2.3 Simplification of equilibrium procedure

This approach concerns ways in which the computation of equilibrium can be simplified. Generally, the objective is to reduce computational expense and computing time
but this does not usually represent a large part of the cost of a study.

Variation in accuracy of computation

Some procedures for computing equilibrium provide the option of sacrificing accuracy
in order to decrease computational cost - e.g., by reducing the number of iterations in
capacity restraint procedures, or increasing the increment size in incremental assignment
procedures.

Use of direct approach

Instead of using a four-step process of the type outlined in section II.3, equilibrium can be computed in a single-step, direct approach, using the explicit forms of demand
models outlined in section II.4.

"Pivot point" approaches*

Assume that the flow pattern (F_1) for one particular set of transportation options
(T_1) has been produced (e.g., as output from an assignment). Now it is proposed to test
a modification of T_1, represented as T_2. "Pivot point" approaches estimate the corresponding flows F_2 by modifying the base flow pattern F_1 in some manner corresponding to
the difference between T_1 and T_2.

For example, if T_2 deletes a link from the network of T_1, then one pivot point
algorithm might take the flow over that link (from all origins to all destinations) and
redistribute it over other links in the network, to yield F_2.

V.2.4 Omission of variables

One obvious way to simplify models is to omit certain variables. For example, in
the case of level of service, instead of a number of variables, ignore cost, frequency of
service, walk distance, etc., and include only travel time. This amounts to assuming
zero elasticities for the omitted service variables.

In some towns it may be possible to omit Public Transport analysis; in others,
car parking analysis may be simplified.

Obviously, omission of variables can lead to serious biases in the results in some
cases, and be perfectly acceptable in others. Each case must be judged on its merits.

V.2.5 Reduced data collection

As outlined in section V.1, an undue amount of time and money is often spent in
the data collection and analysis stages of current models. Reductions in the amount of
data to be collected would, therefore, be of major benefit and would permit more investment in the extraction of information in a form of greater value for transport planning.

Omission of home interview

In a number of countries, research has been carried out into ways for simulating
travel without recourse to home interview surveys for data collection. One method estimates the parameters of "gravity" trip distribution models using traffic volume counts[73].
Another method develops and assigns "trip probabilities" as distinct from "trips" to the
transportation network[74].

---

* See section III.3.4.

The trip probability between two zones i and j is expressed in terms of

$$\frac{A_i B_j}{t_{ij}^{m}}$$

where A and B are parameters that are logically related to trip productions and attractions. For example, for home based work trips A might be population and B employment. The separation factor $1/t_{ij}$ is the time, distance or generalised cost of travel between zones i and j, and m is some exponent value (which may take a different value for different combinations of A and B). A more elaborate formulation may be used for the separation factor $1/t_{ij}^{m}$ if data are available. The trip probabilities are assigned to the network and the resulting "volumes" are related to actual traffic counts at a number of points on the network using multiple regression techniques. If the regression coefficients so obtained can be assumed to be constant with time, then the regression equations can be used to estimate future travel by varying the A and B parameters.

Prespecified parameters

Another obvious way to simplify models is to avoid altogether the problems of data collection and analysis in order to calibrate a model. One way to do this is to use a set of parameters which have been developed elsewhere(71). For example, work in Germany has produced a large volume of traffic survey data. This makes it possible to build up disaggregate models. The basic elements of these models are households with specific behaviour and socio-economic backgrounds. Uni-behavioural households form a group for which prespecified parameters can be fitted. Earlier work has produced a manual(48) for direct calculation of the peak hour traffic generated by urban residential areas.

Another study(1) aimed at determining daily traffic for certain journey purposes in connection with the structure of town cells enabled to obtain first rough estimates to be made for traffic forecasting.

Whether use of prespecified parameters is acceptable depends on the situation. It is important to recognise explicitly the range of uncertainty in the parameters and to study the sensitivity of the forecast volumes to alternate assumptions about the parameters.

V.2.6 Computer environment

Many things can be done to make the models appear simpler to the user and be easier to use in purely mechanical terms. While computing cost and/or time may be increased, total analysis time and cost - including the analyst as well as the computer - may be significantly reduced.

Graphic display

While not affecting the computational cost and time of the model itself, substantial improvements in analyst cost and time expenditures can be achieved by the use of graphic display devices in an on-line environment.

Problem-oriented languages used

Instead of FORTRAN or similar computer-oriented languages, languages tailored closely to the transportation analyst's terminology can reduce analyst time significantly.

V.3   Concluding Remarks

One of the objectives of the Group was to consider the minimum data requirements for an acceptable model both for present and future conditions. This objective presented the Group with a major problem for which no satisfactory solution was found.

As has been mentioned earlier (Chapter IV), each urban traffic model is the result of a balance between resources allocated to model formulation, data collection, model calibration and the testing of alternative transportation strategies. Levels of accuracy acceptable (and consistent with the data and the uses to which the model will be put) enter into each stage, and the whole study process must usually be completed within some maximum time span. Policy decisions are also involved: in some cases a quick and rough analysis may be acceptable. while in others, fine detail may be required.

The descriptions of current work on model simplification being carried out in Member countries were of little assistance to the Group in deciding on minimum data requirements. The presence of items specific to individual areas (see also reference 23) makes comparisons of studies in different countries very difficult - even in cases where enough information about the studies is available. Furthermore, little work has been done in assessing the way in which the overall accuracy of an individual study varies as the data requirements are reduced.

Hence, with so many variables operating, it became clear that the Group could not reach a definitive conclusion on the minimum data required for an acceptable urban traffic model. The best the Group felt it could do was to outline normal data requirements (section V.1), ways in which data could be simplified (section V.2), or have been (section III.1), and accuracy considerations influencing observed (and theoretical) accuracy levels from various studies (chapter IV).

The situation as described is a strong indication that more research is needed on the overall accuracy of transport planning activities and the influence of errors in the components involved.

VI

SUMMARY, CONCLUSIONS AND RECOMMENDATIONS

VI.1   General

As outlined in the Introduction, the Group was given four specific objectives in their Terms of Reference. The first objective was a review of the present stage of development of urban traffic models with particular reference to their complexity and validity. This review is contained mainly in Chapter II. Chapter III examines the features of existing simplified traffic models and ways in which these models might be improved. Chapter IV examines accuracy considerations and Chapter V deals with data requirements and possible simplifications. Each of these items is discussed in relation to examples of existing models of real situations.

The following sections of the Report deal with the Group's conclusions regarding these objectives including the specification of future research needs.

VI.2   Simplified Models

It was agreed that both detailed and simplified models are required for transport planning purposes. It is difficult to define a degree of simplification since all models include simplified representations but the class of model on which the Group concentrated has consisted of those models which were employed for strategic studies of urban development. Such models should deal with various policy effects and should produce results quickly for each new alternative studied. These results should be clearly intelligible to non-specialists. Input data should not demand expensive special-purpose surveys and should use generally available existing sources as far as possible. The interaction between the traffic components of the models and the environmental and land-use characteristics of the area studied is clearly of vital importance and affects the evaluation of the utility of transport as a part of the urban fabric. It is suggested that accessibility could be developed to provide a consistent means of evaluating land use effects (Appendix 4).

A survey of current research in the Member countries indicated two main themes in approaches to modelling. In order to discuss simplification, these themes need to be considered separately.

- The first involved a process of refinement, thinning out or generalisation of the conventional four-step modelling process.

- The second attempted to define a new family of models in which the travel demand and the transport supply are linked in a closed loop which automatically finds its own equilibrium.

The latter type of "equilibrium" model is fundamentally better suited to the study of future changes but can suffer from lack of input data of the required type, increased computing complexity and lack of familiarity for most traffic engineers. However, as traffic restraint, whether by congestion, or by artificially applied means, becomes of increasing importance it may become necessary to adopt equilibrium models for this reason alone. The essential property of equilibrium models is their ability to represent internal and external interactions in a system involving numerous variables.

The importance in policy issues of identifying groups of the population who gain or lose by changes in the transport situation and the effect of the balance between public and private transport on such questions and on infrastructure planning mean that a considerable degree of disaggregation is inevitable in models used for strategic studies. There are thus a number of factors working towards an increase in complexity in such models and means of combating this trend have been an important preoccupation of the discussions in the Group.

VI.3 Four-step Models

In spite of their fundamental limitations, four-step models have been widely used for transport planning purposes. Their adaptation for strategic purposes depends upon the extent to which they can be simplified and still produce useful results, quite apart from the question of whether they are sufficiently responsive to changes in transport conditions. Taking the four steps individually the following points can be made:

(a) Trip* Generation

It has been the practice to derive generation functions for individual towns from home interview surveys. This is an expensive process and there is considerable demand for generalised data for strategic purposes. Variation in these functions between cities can be considerable, even with allowance for such factors as level of car ownership. Study of the stability of generation functions would be welcomed and means by which past household surveys and national census data could be integrated to serve future planning data input needs were rated a high priority by the Group.

Examples of simplifications which have been adopted (see section III.1) are the restriction of journey purpose to two categories or even to one category, the use of a regional rather than a detailed level of analysis, simplified techniques for deriving a future trip matrix from an existing one, and the derivation of average generation factors for suburban residential areas. A method being developed (III.1.3) uses existing travel data for small number of zones and relates the aggregate travel patterns to land use planning without the use of a computer (see also Section III.2.2). ·

The overall criticism which affects all the above examples, however, is that trip generation is frequently not explicitly a function of the level of service provided by the transport system and this can therefore give rise to highly unrealistic conclusions (e.g. failure to predict increases in traffic when new roads are built). Attempts have been made to relate trip generation to level of service (or perhaps accessibility) but this is not a part of the basic model structure and inconsistencies can occur. However the theory referred to in Appendix 1 provides a systematic basis for such developments.

(b) Trip Distribution

The process of estimating the number of trips between pairs of zones has provided little scope for introducing effects of land use and transport system changes. By divorcing distribution from generation the full impact of new developments can be lost, in spite of the apparent advantage of being able to identify, say, the effect of a change in attraction in a given zone. The division of the conventional model into its four steps is the source of its weakness: it tends to lack feed-back from one stage to another. Iterative methods have been used to overcome this difficulty but then

---

* The use of the term trip as a unit of travel was found to be potentially misleading - the associated trip length distribution and definition of when linked trips are regarded as a single trip are typical of the additional information needed to avoid ambiguity.

computing problems escalate sharply without the advantages of a self-consistent model structure.

(c) Modal Split

The division of trips between the modes of transport available can be done before or after the distribution stage. In the former case the effect of transport system efficiency on mode choice is lost and in the latter case movements between zone pairs cannot normally be affected by mode characteristics. In urban situations where important changes in the modes used for certain journeys are likely, these difficulties present serious problems. Most members of the Group are seeking better information on the effect of mode characteristics on patronage, the incorporation of a sound relationship between choice of mode and route, and a balanced way of allowing for the interaction between land-use changes and (public) transport developments. Particular mention is made of the influence of fares, car park and other charges and of the need to ensure that generalised costs are consistent with assignments. In particular circumscribed cases some success has been achieved using tables of present modal split for a few sub-divisions of a town; also very simplified descriptions of level of service have been used to give broad indications for planning purposes. If there is a prospect of a "new" mode being added to the system the four-step process is particularly weak.

(d) Assignment

It is at this stage that the main effect of transport supply on demand is introduced into the four-step process but it is only by unusually powerful feed-back and re-iteration procedures that allowance for this can be made adequate.

Thus the four-step model can only be made really satisfactory by an intimate inter-linking of its stages in a way that brings it close to a genuine equilibrium model. For the purposes of strategic planning this degree of complexity is not usually acceptable and in many cases such planning has employed very elementary components of the four-step process, the validity of which can only be regarded as satisfactory in particular limited circumstances.

VI.4 Demand Supply Equilibrium Models

These models combine the four steps discussed above into a single model structure which includes the necessary feed-back connections and hence computes an equilibrium which satisfies all the functions describing the total transport system and its' users. A variety of such models can be constructed using, for example, the DODOTRANS or RRLTAP systems of computer programmes. The basic theory behind such models and their relationship to the four-step type of model has been summarised in this report. A great deal more needs to be done to develop both the modelling concepts and the necessary input data but there are strong indications that the equilibrium class of model possesses the fundamental inter-linkages which should be characteristic of a tool suitable for broad strategic studies. If the equilibrium type of model is adopted, various ways of simplifying it can be used. One example which has been tried is the ring-radial idealisation of the network employed in CRISTAL. Future developments should include tests of simplification of all the functions included in the model structure so that important additions to the breadth of the factors studied (such as environmental effects) can be made without too unwieldy a result.

VI.5 Computer Aspects

On the subject of computing complexity it should be remembered that computer costs need not be a large component of the cost of model studies so long as data handling (and thereby labour and manpower cost) is kept to a minimum. Much more vital considerations are the ease with which computer models can be constructed to meet particular needs and the facilities for communication between the planning personnel and the computer. The provision of modular programme packages has greatly contributed to the former and prospects for the development of suitable visual presentation and interrogation devices are a line of development which should be pursued to assist in the latter case.

VI.6   Peak-Hour Factors

Nearly all traffic modelling studies a given instant during the day. The inclusion of time history effects to permit the modelling of such effects as peak spreading would involve a great increase in complexity and the alternative of using an empirical factor to relate peak hour traffic to daily traffic is usually adopted. The Group recognised that more work is needed on the time period cross-elasticity of traffic with special attention to possible policy interventions such as staggered working hours. This work could have important implications for the validity of traffic forecasts. Although this introduces a new dimension of complexity, developers of simplified traffic models should consider explicit inclusion of time period cross-elasticities or alternatively look for methods which provide peak hour conditions directly. It should be mentioned that in many cases networks cannot be designed to meet peak loads and the resulting effects in peak spreading become a major interest.

VI.7   Land Use

It is important for strategic studies that the interaction between transport and land use should be represented. if only in simplified terms. For traffic planning purposes, it has commonly been the practice to take a land-use plan as an input and hence estimate the traffic generated; this can be misleading in the case of important changes in the transport system.

An example of the type of model which assists in the understanding of land-use/ transport interaction uses an accessibility index which combines the various choices offered by a city with the ease of access to them in a function which can be evaluated on a zone-zone basis. Types of employment and recreation can be treated separately or combined to built up a picture of service provided and traffic resulting. It should be mentioned, however, that such traffic would not be subject to capacity constraints in the simple type of model envisaged.

Practical cases demand the minimising of the number of variables to be considered. Typically an urban area might be described by its population, number of employed, number of employment opportunities and percentage of those which are tertiary. Comparison between results obtained from such a simplified picture and those from a full household survey show that the weighted error affecting traffic generation for the simpler model amounts to 10-20 per cent. This might be acceptable for broad strategic purposes.

VI.8   Public Transport

Urban transport planning is almost universally concerned at present with the need to transfer traffic from private cars to public transport, which may be of a novel form. The modelling of this process is thus a vital factor.

Estimation of future patronage from tables of current modal split can be used to give a rough estimate for strategic purposes but methods which include the effect of level of service are recommended wherever possible. A "pivot point" approach using an elasticity to a measure of level of service such as frequency can be simple and easy to apply.

The use of more complete generalised cost functions to describe public transport journeys can reflect the combined effects of a number of interacting parameters but tend to increase model complexity. Ways of minimising this difficulty include:

(a) use of crow-fly distances;
(b) use of zone average speeds;
(c) combination of many trip purposes.

The choice of simplification must depend upon the purpose for which the model is to be used, since for some strategic studies disaggregation (e.g. of trip purpose) is essential. Thus there is a need for a range of models for use by transport planners and the choice of model for each individual purpose is an area of work which has not yet been fully explored.

Where the introduction of a new mode of transport is to be studied demand equilibrium models are much preferable to the four-step process since the reaction between supply and demand is a dominant feature of this situation.

The need for public transport to meet the needs of particular sectors of the population (e.g. young, old, infirm, etc.) can override simple economic considerations and must be allowed for in the situation modelled.

VI.9   Accuracy Considerations

A transportation study has to make a compromise between a wealth of detailed data on the one hand and a sophisticated and complex model on the other. With the usual limitations on total resources available, allocations to one aspect mean loss to the other and guidance on the correct compromise is not readily available. It is possible to relate percentage errors to size of sample in data collection but very little can yet be said about the relationship between model features and the accuracy of analysis which it provides. Examples are now emerging where model results specifically include statements of computation errors but at present these are the exception rather than the rule.

The problem is compounded by the fact that strategic planning frequently requires results in disaggregated form. However, a judicious choice of level of aggregation to meet particular needs can have a beneficial effect on both data accuracy and model complexity. Usually disaggregated data are derived from a wide cross-section of the population which is classified according to socio-economic groups or accessibility factors. The concept of disaggregate models in which individual behaviour is represented by a probability distribution promises well for valid models based on small data samples. Further work on development of such models should be encouraged. The need for accuracy statements at all stages of transport studies and the desirability of sensitivity tests cannot be over-emphasized. Simplified models greatly facilitate the provision of this sort of information and the need for their development could be based upon this feature alone.

VI.10   Recommendations

The leading recommendations of the Group are summarised in the following items:

1. Simplified models facilitate the rapid analysis of many alternative transport policies and help the study of the sensitivity of forecasts to sources of uncertainty such as model parameters and external assumptions. Some models of this type exist and have been described in this report. They could be employed in some cases but considerable development is still desirable.

2. Equilibrium models provide the internal interactions necessary for studies in cases where important changes in the transport situation can be anticipated. Conventional four-step models are not well suited to such problems.

3. Where four-step models are employed for strategic studies it is desirable to integrate the computer programs so that the user can run the entire model in a single step. This facilitates rapid analysis and offers the possibility of using iterations to study changing conditions while maintaining consistency. Ordinary four-step models based on households can be simplified by using few types of household and fewer travel purposes. Household models are rather complicated and require home interview data. A new type of model based on traffic counts offers simplicity and results can serve strategic planning purposes.

4. It is essential that errors involved in each of the stages of the data collection and modelling processes are clearly stated and the trade-off in the allocation of resources should be understood.

5. There is considerable demand for standardisation of definitions and format in data collection so that data banks can be accumulated and accessed for many different purposes. This is not solely related to simplified models but the possibility of international co-operation in this field should be studied.

6. Since disaggregate modelling techniques are particularly promising for simpli-fied models, priority should be given to the international exchange of data suitable for their estimation.

VI.11 Further Research Requirements

A number of research items of special importance for the development of effective strategic modelling capability have been highlighted during the work of the Group. Individual items on which research and liaison should be encouraged are listed below; the order is the sequence of discussion above rather than an order of priority:

(i) The development of simplified models designed specifically for strategic transport planning purposes should be recognised as having special require-ments. A number of such models are beginning to appear and liaison to dis-seminate information on this work should be established.

(ii) Development of criteria describing the utility of transport as part of the urban structure. This would hopefully lead to a new definition of the benefit derived by the community from transport developments.

(iii) A component of (ii) is a means of including environmental factors in transport system evaluation. Currently a great deal of work is being done on this aspect but means of condensing detailed studies into forms suitable for application in strategic models will require special attention.

(iv) Demand equilibrium models have been advanced as a way to improve the handling of the interaction effects which become important when exploring future con-ditions. More work is needed to elucidate the relationship between such models and the more conventional transportation models. Also the possibilities for reducing data input requirements and computation time for equilibrium models have not yet been fully exploited. Theory indicates that much could be gained in this direction.

(v) In conventional models there have been many attempts to produce generalised trip generation functions to obviate the need for expensive special surveys.

These have met with varying success but there is clearly a case for a comprehensive study for European conditions employing existing survey data. Such a project would produce results of value throughout transportation planning activities. If level of service of transport can be included as a parameter, this work would be especially worthwhile.

(vi) More information on the effect of mode characteristics on patronage with the effect of routing, land-use changes and other parameters explicitly included was a unanimous requirement. With so many cities looking to public transport enhancement to overcome their congestion problems, this was naturally a topic of urgent interest. However this is not an item which is special to the development of simplified traffic models.

(vii) Studies of the elasticity of various categories of traffic to generalised cost of travel could include mode choice factors and provide vital basic information on travel behaviour on which models of future traffic conditions are based. The results clearly depend upon the local conditions and a variety of results can be expected. The continued association of the Group could provide a forum in which the results of experiments and studies in this field could be compared.

(viii) Time elasticity is a special case of (vii) and special studies of peak hour spreading and the implications for traffic forecasts are required.

REFERENCES

1. Riemer,P.M., "Entwicklung von Standard-Verkehrsplänen einschliesslich ihrer Elemente und Lösungsvorschläge für Verkehrsanalysen und -prognosen einzelner Stadt-Grössenklassen," Schriftenreihe "Forschung Stadtverkehr," Volume 4, Bundesminister für Verkehr, Bonn, 1970.

2. Voigt,F. "Welche Grundsätze sind bei der Generalverkehrsplanung der Städte, der Regionen, der Länder und des Bundes zu beachten, um ein einheitliches und koordiniertes Verkehrssystem in der Bundesrepublik zu erhalten und zu entwickeln?" Schriftenreihe "Forschung Stadtverkehr," Volume 2, Bundesminister für Verkehr, Bonn, 1967.

3. O.E.C.D., "Effects of traffic and roads on the environment in urban areas," Paris, 1973.

4. Ruske,W. "Entwicklung der Vorstellungen zum Verkehrserzeugungsmodell für die verschiedenen Planungsebenen," Schriftenreihe des Institutes für Stadtbauwesen an der Rheinisch-Westfälischen Technischen Hochschule Aachen, Volume 14, Aachen, 1970.

5. Kutter, E. "Demographische Determinanten städtischen Personenverkehrs," Veröffentlichung des Institutes für Stadtbauwesen an der Technischen Universität Braunschweig, Volume 9, Braunschweig, 1972.

6. Ruske, W., Stelling, H., "Wertung und Weiterentwicklung der als Grundlage für Strassenverkehrsplanungen dienenden Verkehrserhebungsmethoden," Schriftenreihe "Strassenbau und Strassenverkehrstecknik" des Bundesministers für Verkehr, Abt. Strassenbau, Volume 79, Bonn, 1968.

7. Manheim, M.L., "Practical implications of some fundamental properties of travel demand models", Highway Research Record 422. Wash. D.C.

8. Brand, D., and Manheim, M.L., "Directions for research in travel forecasting", Proceedings of the First International Conference on Transportation Research, Transportation Research Forum, Chicago, Illinois, (in press).

9. Martin, B.V., Memmott, F.W. and Bone, A.J., "Principles and techniques of predicting future urban area transportation," Cambridge, Massachusetts: M.I.T. Press, 1965.

10. Mäcke,P.A., "Analyse- und Prognosemethoden des regionalen Verkehrs," Schriftenreihe des Institutes für Stadtbauwesen an der Rheinisch-Westfälischen Technischen Hochschule Aachen, Volume 6, Aachen, 1968.

11. Systems Analysis and Research Corp., "Demand for intercity passenger travel in the Washington-Boston corridor," U.S. Department of Commerce, 1963.

12. Quandt, R.E. and Baumol, W.J., "Abstract mode model: theory and measurement," Journal of regional science, Volume 6, No. 2, 1966.

13. Kraft, G. and Wohl, M. "New directions for passenger demand analysis and forecasting." Transportation Research, Volume 1, 1967.

14. McLynn, J.M. et al, "A family of demand modal split models," Arthur Young and Company, April 1969.

15. McLynn, J.M. and Watkins, R.H., "Multimode assignment model," Washington, D.C., National Bureau of Standards, 1965.

16. Plourde, R. "Consumer preference and the abstract mode model: Boston metropolitan area," Research Report R68 51, Cambridge, Massachusetts: Department of Civil Engineering, M.I.T., 1968.

17. Domencich, T.A., Kraft, G. and Vallette, J-P, "Estimation of urban passenger travel behaviour: an economic demand model" Transportation system evaluation, Highway Research Record 238, Washington, D.C.: Highway Research Board, 1968.

18. Wigan, M.R. "Benefit assessment for network traffic models and application to road pricing," RRL Report 417, 1971.

19. Tanner, J.C., Gyenes, L., Lynam, D.A., Magee, S., Tulpule, A.H., "The development and calibration of the CRISTAL transport planning model," TRRL Report 574, 1973.

20. Ben-Akiva, E., "Structure of Passenger Travel Demand Models", unpublished PhD thesis, Cambridge, Massachusetts: M.I.T. Department of Civil Engineering, 1973.

21. Ben-Akiva, E., "A Disaggregate Direct Demand Model for Simultaneous Choice of Mode and Destination" to be published in Proceedings - International Conference on Transportation Research, Bruges, Belgium.

22. Ben-Akiva, E., "Multi-Dimensional Choice Models: Alternative Structures of Travel Demand Models," to be published in Proceedings of Engineering Foundation Conference held at Berwick, Maine, July 1973.

23. Curran and Stegmaier, "Travel Patterns in 50 cities," Public Roads, N. 5, 1958.

24. Kirchhoff, P., "Verkehrsverteilung mit Hilfe eines Systems bilinearer Gleichungen," Schriftenreihe des Institutes für Stadtbauwesen, Technische Universität Braunschweig, 1970.

25. Meyer, L. "Abschätzung des Verkehrsaufkommens im öffentlichen Personennahverkehr in Wohngebieten," Schriftenreihe "Strassenbau und Strassenverkehrstechnik" des Bundesministers für Verkehr, Abt. Strassenbau, Volume 120, Bonn, 1971.

26. Dornier-System GmbH, "Entwicklung und Programmierung eines Verfahrens zur Suche alternativer Routen im öffentlichen und individuellen Nahverkehr, Schriftenreihe "Forschung Stadtverkehr," Bundesminister für Verkehr, Volume 5, Bonn, 1970.

27. von Falkenhausen, H., "Ein stochastisches Modell zur Verkehrsumlegung" Schriftenreihe "Strassenbau und Strassenverkehrstechnik" des Bundesministers für Verkehr, Abt. Strassenbau, Volume 64, Bonn, 1967.

28. Ueberschaer, M., "Die Aufteilung der Verkehrsströme auf verschiedene Fahrtwege (Routen) in Stadtstrassennetzen aufgrund der Strassen- und Verkehrsbedingungen beim morgendlichen Berufspendelverkehr" Schriftenreihe "Strassenbau und Strassenverkehrstechnik" des Bundesministers für Verkehr, Abt. Strassenbau, Volume 85, Bonn, 1969.

29. Manheim, M.L., and Ruiter, E.R., "DODOTRANS I: A decision-oriented computer language for analysis of multi-mode transportation systems," Costs and benefits of transportation planning, Highway Research Record 314, Washington, D.C., Highway Research Board, 1970.

30. Wigan, M.R. and Bamford, T.J.G., "A perturbation model for congested and overloaded transportation networks," TRRL Report LR 411, 1971.

31. Wigan, M.R., Webster, F.V., Oldfield and Bamford, T.J.G., "Methods of evaluation of traffic restraint techniques," O.E.C.D., Symposium on techniques of improving urban conditions by restraint of road traffic, Paris, 1973.

32. Manheim, M.L., "Fundamentals of Transportation Systems Analysis", (Preliminary Edition) Department of Vicil Engineering, M.I.T., Cambridge, Mass., 1971.

33. Pratt, R.H., and Deen, T., "Estimation of sub-modal split within the transit mode", Highway Research Record 205, Washington D.C.

34. Wegener, M., and Meise J., "Stadtentwicklungssimulation", Stadtbauwelt, Heft 29, 1971

35. Harloff, G., "Ein Optimierungsansatz zur Ordnung der Stadtinhalte, Seminarbericht Nr. 7 der Gesellschaft für Regionalforschung", Innsbruck, February 1973, and "Ordnung der Flächennutzung nack Massgabe der allgemeinen Entwicklung und der Bedingungen des Verkehrs", Dissertation, Aachen, to be published 1974.

36. Institut für Orts-, Regional- und Landesplanung der ETH Zürich, "ORL - MOD 1 - Ein Modell zur regionalen Allokation von Aktivitäten", Arbeitsberichte No. 24.1 1971.

37. Forrester, J.W., "Urban Dynamics", M.I.T. Press, Cambridge (Mass.) 1969.

38. ZBZ, (Zentrum Berlin für Zukunftsforschung), "Das Berliner Simulationsprogramm BESI im Grundriss" analysen und prognosen, Heft 9, May 1970, and "Entwurf eines kommunalen Management-Systems", ZBZ - Bericht 9, Sept. 1970.

39. Fleet, C.R., and Robertson, S.R., "Trip generation in the transportation planning process", Highway Research Record 240, Washington D.C. Highway Research Board, 1968.

40. Kessel, P., "Motivationen für charakteristische Verhaltensformen und mögliche Entwicklungstendenzen im Stadtverkehr", Schriftenreihe des Institutes für Stadtbauwesen an der Rheinisch-Westfälischen Technischen Hochschule Aachen, Volume 20, Aachen, 1970.

41. Mäcke, P.A., Jürgensen, H., "Das Verkehrsaufkommen in Abhängigkeit von der Siedlungs-, Wirtschafts- und Sozialstruktur (Flächennutzung)", Schriftenreihe "Forschung Stadtverkehr", Bundesminister für Verkehr, Volume 1, Bonn, 1967.

42. Rucker, A., Schöpf, G., Springer, O., "Verkehrsaufkommen des ungebundenen Verkehrs von Stadtrandsiedlungen", Schriftenreihe "Strassenbau und Strassenverkehrstechnik" des Bundesministers für Verkehr, Abt. Strassenbau, Volume 43, Bonn, 1966.

43. Drangmeister, K., "Das Kfz-Verkehrsaufkommen von Wohngebieten unter besonderer Berücksightigung der Spitzenverkehrszeiten", Schriftenreihe "Strassenbau und Strassenverkehrstechnik" des Bundesministers für Verkehr, Abt. Strassenbau, Volume 64, Bonn, 1967.

44. Scholz, G., Wolff, G., Heilsch, H., and Schreiber, A., "Untersuchungen über die Struktur und Nutzung des Verkehrsraumes in städtischen Verkehrsnetzen", Schriftenreihe "Strassenbau und Strassenverkehrstechnik" des Bundesministers für Verkehr, Abt. Strassenbau, Volume 124, Bonn, 1971.

45. Kessel, P., "Verhaltensweisen im werktäglichen Personennahverkehr", Schriftenreihe "Strassenbau und Strassenverkehrstechnik" des Bundesministers für Verkehr, Abt. Strassenbau, Volume 132, Bonn, 1972.

46. National reports prepared by Members of the Group:

      (a) Belgium
      (b) Ireland
      (c) United Kingdom
      (d) Finland
      (e) Denmark
      (f) Canada
      (g) Spain

47. Dublin Transportation Study. Technical Report No. 16, paragraph 1625, An Foras Forbartha, Dublin, 1973.

48. Forschungsgesellschaft für das Strassenwesen e.V., "Merkblatt für die Vorausschätzung des Verkehrsaufkommens von städtischen Wohnsiedlungen," Cologne, 1969.

49. SETRA, Notice d'utilisation du programme FABER, Bagneux, 1972.

50. Schneider, J.B., Gehner, C.D., Porter, D., "Man - Computer Synergism: A novel approach to the design of multi-objective, multi-modal urban transportation systems," Paper presented at the International Conference on Transportation Research, Bruges, June 1973.

51. Gaudefroy, O., Demombynes, A., "Le simulateur de déplacements en milieu urbain," Revue Générale des Routes et des Aérodromes, No. 487, May 1973.

52. O.E.C.D., "Optimisation of bus operation in urban areas," Paris, 1972.

53. SELNEC Transportation Study Report, Town Hall, Manchester 1972.

54. Bendtsen, P.H., "Goals for the development of the city centre", International Federation for Housing and Planning, Standing Committee on traffic problems, 1973.

55. Institute of Traffic Engineers, "Change-of-mode parking: A State of the art", January 1973.

56. Report nr 1, Stockholms läns landsting: TU 71, Traffic Surveys in the Stockholm region, 1971.

57. Stockholm General Planning Office: Traffic Programme for Stockholm, A Paper for Discussion, May 1973.

58. Johnson, K.R., "A pedestrian model based on a home interview survey," Planning and Transportation Research Co. Ltd. Conference 1972, London, 1971.

59. Edwards, J.A. and Shipley, S., "The Coventry transportation study walk model," Planning and Transportation Research Co. Ltd.,Conference, 1972

60. Purcivall and Sandahl, J., "Pedestrian traffic forecast model for town centres," Planning and Transportation Research Co. Ltd., Conference 1972, London,1971.

61. Brokke and Mertz, "Evaluating trip forecast methods," Public Roads, October 1958.

62. Crawford, "Accuracy of the transportation model," Greater London Council Transportation, Branch. Res. Memorandum, 1968.

63. Strand, "The stepwise planning procedure," Traffic Quarterly, January 1972.

64. Sosslau and Brokke, "Appraisal of O-D survey sample size," Publ. Roads, December 1960.

65. Lewis, "Trip generation techniques," Traffic Eng. and Control, November 1970, February 1971.

66. Highway Research Board, "Simplified procedures for determining travel patterns", Highway Research Record, 88, 1965.

67. Nordqvist, S. "Studies in traffic generetics," National Swedish Building Research, Doc. no. 2, 1969.

68. Harper and Edwards, "Generation of person trips by areas within the central business district," Highway Research Board Bulletin 253, 1960.

69. Bates, "Development and testing of synthetic generation models for urban transportation studies", State Highway Department, Georgia, PB 206090.

70. Kirsch, H. "Kritischer Vergleich verschiedener Verteilungsmodelle von Verkehrsaufkommen für den Planungsraum, ihre Anwendungsbereiche und ihre Fehleranfälligkeit", Schriftenreihe "Strassenbau und Strassenverkehrstechnik" des Bundesministers für Verkehr, Abt. Strassenbau, Volume 119, Bonn, 1972.

71. Mäcke, P.A., Hölsken, D., "Generalverkehrsplan Ruhrgebeit, Individualverkehr", Hefte 11, 12 der Schriftenreihe des Siedlungsverbandes Ruhrkohlenbezirk, Essen.

72. Irwin, Dodd, Cube, "Capacity restraint in assignment program", Highway Research Board Bulletin 297, 1961.

73. Jensen, T., Nielsens, S.K., "Calibrating gravity models and estimating its parameter using traffic volume counts," Technical University of Denmark, January 1973.

74. Low, D.E., "A new approach to transportation systems modelling," Traffic Quarterly, July 1972, pp. 391-404.

75. Charles River Associates, "A disaggregate behavioural model of urban travel demand", Washington D.C.: Federal Highway Administration, 1972.

76. Reichman, S., and Stopher, P.R., "Disaggregate stochastic models of travel mode choice". Highway Research Record 369, Washington D.C.: Highway Research Board, 1971.

77. Stopher, P.R., and Liscok, T.E., "Modelling travel demand: A disaggregate behavioural approach - issues and application", Transportation Research Forum Proceedings. Oxford, Indiana; Richard B. Cross Co., 1970.

78. SETRA, Notice d'utilisation and Système d'utilisation du programme ATLANTE, Bagneux, June 1974.

79. Wilson, N.H.M., Pecknold, W.M., and Kullman, B.C. "Service modification procedures for MBTA local bus operations", Boston, Mass.: Boston Urban Observatory, Boston College, 1972.

80. Cambridge Systematics, Inc., "Introduction to urban travel forecasting", prepared for U.S. Urban Mass Transportation Administration Software Systems Development Program, Cambridge, Mass: Cambridge Systematics, 1973.

81. Bendtsen, P.H., "Traffic generation", Socio. Econ. Plan. Sciences, 1967.

82. Crawford, "Accuracy of the transportation model". Greater London Council Transportation Research, Research Memorandum, 1968.

83. Douglas and Lewis, "Trip generation techniques", Traffic Eng. and Control, 1971.

84. Sosslau and Brokke, "Appraisal of O-D survey sample size", Public Roads, Dec. 1960.

85. Ben, Bouchard and Sweet, "An evaluation of simplified procedures for determining travel patterns in a small urban area". Highway Research Record No. 88, 1965.

86. Brokke and Mertz, "Evaluation trip forecast methods", Public Roads, October 1958.

87. Zuberbühler, "Technische Aspekte städtischer Verkehrsplanungen", Automobilismo e Automobilismo Industriale, No. 5-6, 1971.

88. Wigan, M.R. and Bamford, T.J.G., "An equilibrium model of bus and car travel over a road network", T.R.R.L. Report LR 559.

89. Webster, F.V. and Oldfield, R.H., "A theoretical study of bus and car travel in central London", T.R.R.L. Report LR 451.

90. Wigan, M.R. and Bamford, T.J.G., "A suite of programs for transportation planning on the ICL 4/70", Proceedings 1970 computer program review Symposium Paris 1970, London 1971. (PTRC)

91. Wigan, M.R. and Bamford T.J.G., "The effects of network structure on the benefits derivable from road pricing", T.R.R.L. Report LR 557.

92. Wigan, M.R. and Bamford, T.J.G., "A comparative network simulation of different methods of traffic restraint", T.R.R.L. Report LR 556.

APPENDICES

APPENDIX 1

GENERAL SHARE MODEL

The General Share Model (GSM) is defined as:

$$V_{klmp} = \alpha(Y)\, \beta_k(Y)\, \gamma_{kl}(Y)\, \delta_{klm}(Y)\, \omega_{klmp}(Y)$$

where:

$$Y = f(R,Z) = f(\underline{A}, \underline{a}, \underline{X}, \underline{w})$$

where α, β, γ, δ, ω are functions which meet the following Range Conditions for all values of Y:

$$0 \leqslant \alpha(Y);$$

$$0 \leqslant \beta_k \leqslant 1, \sum_k \beta_k(Y) = 1;$$

$$0 \leqslant \gamma_{kl}(Y) \leqslant 1, \sum_l \gamma_{kl}(Y) = 1 \text{ for every } k;$$

$$0 \leqslant \delta_{klm}(Y) \leqslant 1, \sum_m \delta_{klm}(Y) = 1 \text{ for every } k,l;$$

$$0 \leqslant \omega_{klmp}(Y) \leqslant 1, \sum_p \omega_{klmp}(Y) = 1 \text{ for every } k,l,m.$$

The Basic variables are:

V_T = total volume of trips (interzonal) in the region = $\alpha(Y)$

V_k = trip generation = total volume originating in zone k = $\beta_k(Y)\, V_T(Y)$

V_{kl} = trip distribution = volume going from zone k to zone l = $\gamma_{kl}(Y)\, V_k(Y)$

V_{klm} = modal split = volume going from zone k to zone l by mode m = $\delta_{klm}(Y)\, V_{kl}(Y)$

V_{klmp} = trip assignment = volume of trips going from zone k to zone l by mode m and path p = $\omega_{klmp}(Y)\, V_{klm}(Y)$

$\underline{A}$ = vector of variables describing the socio-economic activity system

$\underline{a}$ = vector of parameters applying to $\underline{A}$

$\underline{X}_{klmp}$ = vector of S level of service variables (i = 1, 2, . . . S) describing the transportation system characteristics as experienced by trips from k to l by mode m and path p

$\underline{X}$ = $\underline{X}_{klmp}$ = set of all level of service characteristics for all paths of p of all modes in between all origins k and all destinations l

$\underline{X}'$ = vector of parameters applying to $\underline{X}$

R_{klmp} = $f(\underline{X}, \underline{w})$ = combined effect of all level of service characteristics of all modes as they influence trips from k to l by mode m and path p - a "generalised cost"

$R_{klmq,p}$ = combined effect of all level of service characteristics of mode q as they influence trips from k to l by mode m and path p

Z = $f(\underline{A}, \underline{a})$ = combined effect of all activity system characteristics

Y = $f(Z,R)$ = combined effect of all activity system and level of service characteristics

Table 1 presents some forms of demand models and Table 2 includes forms of the model type SPM-3.

Several basic properties of the GSM have been demonstrated(7)*:

- any <u>explicit</u> model, of the form $V = \Psi(Y)$
  can be written as an equivalent GSM;

- any <u>sequential implicit</u> model, of the form:

$$
V \begin{cases}
V_k & = \sigma_1(Y) \\
V_{kl} & = \sigma_2(Y, V_k) \\
V_{klm} & = \sigma_3(Y, V_{kl}) \\
V_{klmp} & = \sigma_4(Y, V_{klm})
\end{cases}
$$

  can be written as an equivalent GSM
  (provided it is internally consistent)(7);

- the GSM can be written either as an explicit form as given above or as an
  equivalent internally-consistent sequential implicit form:

$$
\begin{aligned}
V_T & = \alpha(Y) \\
V_k & = \beta_k(Y) \cdot V_T \\
V_{kl} & = \gamma_{kl}(Y) \cdot V_k \\
V_{klm} & = \delta_{klm}(Y) \cdot V_{kl} \\
V_{klmp} & = \omega_{klmp}(Y) \cdot V_{klm}
\end{aligned}
$$

Thus, as a consequence of the above, <u>any</u> explicit demand model can be written equivalently
as a sequential implicit model, and any internally-consistent sequential implicit model
can be written equivalently as an explicit model.

---

* See list of references of main report.

TABLE 1: SOME FORMS OF DEMAND MODELS

MODEL TYPE & EQUATIONS	TOTAL TRIPS	GENERATION	DISTRIBUTION	MODAL SPLIT
GSM	$V_T = \alpha$	$V_k = V_T \cdot \beta_k$	$V_{kl} = V_k \cdot \gamma_{kl}$	$V_{klm} = V_{kl} \cdot \delta_{klm}$
SPM-1	$V_T = \{\sum_k Z_k [\sum_l Z_l R_{kl.}^{\delta_{12}}]^{\delta_{21}}\}^{\delta_{30}}$	$V_k = V_T \cdot \left\{ \dfrac{Z_k [\sum_l Z_l R_{kl.}^{\delta_{11}}]^{\delta_{20}}}{\sum_k Z_k [\sum_l Z_l R_{kl.}^{\delta_{11}}]^{\delta_{20}}} \right\}$	$V_{kl} = V_k \cdot \left\{ \dfrac{Z_l R_{kl.}^{\delta_{10}}}{\sum_l Z_l R_{kl.}^{\delta_{10}}} \right\}$	$V_{klm} = V_{kl} \cdot \left\{ \dfrac{R_{klm}}{R_{kl.}} \right\}$
SPM-2	$V_T = \{\sum_k Z_k [\sum_l Z_l R_{kl.}^{\delta_1}]^{\delta_2}\}^{\delta_3}$	$V_k = V_T \cdot \left\{ \dfrac{Z_k [\sum_l Z_l R_{kl.}^{\delta_1}]^{\delta_2}}{\sum_k Z_k [\sum_l Z_l R_{kl.}^{\delta_1}]^{\delta_2}} \right\}$	$V_{kl} = V_k \cdot \left\{ \dfrac{Z_l R_{kl.}^{\delta_1}}{\sum_l Z_l R_{kl.}^{\delta_1}} \right\}$	$V_{klm} = V_{kl} \cdot \left\{ \dfrac{R_{klm}}{R_{kl.}} \right\}$
SPM-3	$V_T = \{\sum_k \sum_l Z_k Z_l R_{kl.}\}$	$V_k = V_T \cdot \left\{ \dfrac{Z_k \sum_l Z_l R_{kl.}}{\sum_k \sum_l Z_k Z_l R_{kl.}} \right\}$	$V_{kl} = V_k \cdot \left\{ \dfrac{Z_l R_{kl.}}{\sum_l Z_l R_{kl.}} \right\}$	$V_{klm} = V_{kl} \cdot \left\{ \dfrac{R_{klm}}{R_{kl.}} \right\}$
CONVENTIONAL FOUR-STEP MODELS	$V_T = \sum_k Z_k$	$V_k = Z_k$	$V_{kl} = V_k \cdot \left\{ \dfrac{Z_l R_{kl.}'}{\sum_l Z_l R_{kl.}'} \right\}$	$V_{klm} = V_{kl} \cdot \left\{ \dfrac{R_{klm}''}{R_{kl.}''} \right\}$

94

TABLE 2

FORMS OF SPM-3

GENERAL FORM: $V_{klm} = $ $\qquad Z_k \qquad Z_l \qquad R_{klm}$

SPECIFIC FORMS:	Z_k	Z_l	R_{klm}
GRAVITY MODEL	P_k	E_l	t_{kl}^{-2}
KRAFT-SARC	$a_m P_l^{e_m} Y_l^{g_m}$	$P_l^{e_m} Y_l^{g_m}$	$\begin{pmatrix} b_{mm} & d_{mm} \\ c_{klm} & t_{klm} \end{pmatrix} \begin{pmatrix} b_{mn} & d_{mn} \\ c_{kln} & t_{kln} \end{pmatrix}$
KDV-1	$a_m P_k Y_k^{g_m}$ $(P_k = H_k^{e'_m} N_k^{e''_m})$	$Q_l^{h_m} S_l^{f_m} T_l^{i_m}$	$\begin{pmatrix} b_{mm} & d_{mm} \\ c_{klm} & t_{klm} \end{pmatrix} \begin{pmatrix} b_{mn} & d_{mn} \\ c_{klm} & t_{klm} \end{pmatrix}$

Additional variables:

P_k	=	population at zone k
H_k	=	number of households at zone k
N_k	=	number of persons per household in zone k
Y_k	=	median income of zone k
E_l	=	employment at zone l
Q_l	=	density of employment in retail trade in zone l
S_l	=	employment in retail trade in zone l as a proportion of total employment in retail trade in region
T_l	=	employment in personal business activities in zone l as a proportion of total employment in retail trade in personal business in region
e_m, f_m, g_m, h_m, i_m	=	parameters for activity system variables
b_{mm}	=	direct elasticity effect of mode m's cost on mode m
b_{mn}	=	cross-elasticity effect of mode n's cost on mode m
d_{mm}	=	direct elasticity effect of mode m's time on mode m
d_{mn}	=	cross-elasticity effect of mode n's time on mode m

Domencick, Kraft and Vallette[1] recently calibrated a set of models for urban passenger trips. The form shown in Table 2 is for transit shopping trips. (Similar models, though somewhat more complex in form, were calibrated for work and shopping trips by car, and public transport work trips.) The estimated values of the parameters were:

$$a_m = -2.0 \qquad g_m = -0.05 \qquad b_{mm} = -0.6$$
$$e'_m = 1^* \qquad h_m = 0.03 \qquad d_{mm} = -0.3$$
$$e''_m = 2.5 \qquad f_m = 1.0^* \qquad b_{mn} = 0$$
$$i_m = -0.74 \qquad d_{mn} = 0$$

(1) Reference 17 of the main report.

APPENDIX 2

EXAMPLES OF DISTRIBUTION MODEL FORMULAE

A type of distribution model which has been used in many studies is based on the so-called Gravity Model formula:

$$t_{ij} = p(i)\ a(j)\ f(i,j) = \text{trips between zones } i \text{ and } j$$

where $p(i) = \dfrac{P_i}{\sum\limits_{j} a(j)\ f(i,j)}$

with P_i = the trips produced in zone i

$a(j)$ = the trips attracted to zone j

$$f(i,j) = k_{ij}\ e^{\theta d_{ij}}$$

where k_{ij} = interzonal "adjustment" factor

d_{ij} = travel time, distance or generalised cost between zones i and j

θ = a(negative) calibration parameter

alternatively $f(i,j) = k_{ij}(d_{ij})^{\alpha}$

where α = a (negative) calibration parameter,

or, a Table giving the value of $f(i,j)$ for each value of d_{ij} might be used.

A modification of the Gravity Model formula has been used in a number of studies in the United Kingdom. It involves the imposition of an additional constraint at the attraction end, thus causing the model to be named a Doubly Constrained Gravity Model. The additional constraint imposed is:

$$a(j) = \dfrac{A_j}{\sum\limits_{i} p(i)\ f(i,j)}$$

with A_j = the trips attracted to zone j

Data collected in the home interview survey are used for estimating the value of the calibration parameters α, θ and k_{ij}.

Some research has been carried out to see if a relationship can be found between mean travel time, value of α and size of city, thus making it possible to do without a home interview survey for distribution model calibration[1,2]. The research indicates that for cities up to about 700,000 inhabitants, mean travel time seems to increase with city size, and might depend on average speed on the city streets. For the cities beyond 700,000 inhabitants dealt with in [1,2] below, mean travel time lies between 13 and 20 minutes. However, a relationship has not yet been established.

The following data were obtained in France:

City	Average travel time (mins.)							
	Home - work				Home - other destinations			
	Private car	Public transport	2-wheelers	Walk	Private car	Public transport	2-wheelers	Walk
PARIS 8003000	25	44	16	14	24	38	14	16
MARSEILLE 915000	24	35	21	17	23	37	17	14
LYON 835000	19	31	19	16	20	31	19	14
BORDEAUX 493000	15	27	17	13	15	25	13	12
ROUEN 348000	19	32	18	16	17	32	17	13
ST. ETIENNE 345000	14	25	14	14	13	26	12	11
NICE 307400	17	27	18	16	19	29	14	15
GRENOBLE 281000	19	27	18	16	17	26	15	14
NANCY 224500	15	27	16	16	13	21	15	13
ORLEANS 162000	17	26	18	15	15	25	15	13
BESANCON 117000	12	22	13	13	11	24	12	12
CHERBOURG 78600	15	27	17	17	13	22	14	13
AIX-en-PROV. 72700	16	21	19	16	15	22	16	15
CHAMBERY 64400	14	18	17	14	14	20	14	14
ELBEUF 45700	14	25	16	15	11	23	12	14
DIEPPE 38200	13	22	15	17	16	20	13	14

References

1. Kirby, "Interim Review of the Results of Gravity Model Calibrations", PTRC Seminar on Urban Traffic Model Research, 1970.

2. Voorhees, "Alternative Land Use and Transportation Policies", Congress of International Federation for Housing and Planning, Liverpool, 1972.

ABSTRACTS OF THREE SIMPLIFIED TRAFFIC MODELS BASED ON TRAFFIC COUNTS

The home interview is the most expensive part of the work with usual traffic models. Some authors have suggested that models can be based on results of traffic counts. Such counts can be carried out without great cost.

Models described below are based on traffic counts and calculate the number of car trips; modal split is not involved. They dispense with a "home interview" survey.

I.   Abstract of:

Low: Transportation system modelling. Traffic Quarterly. July 1972

The model calculates the traffic in a city.

1. Determining the external traffic

Roadside interviews held on the streets entering the city indicated the external traffic to each of the city's zones. This gave a matrix for the external traffic. This desire line traffic was assigned to the city streets.

2. Determining the travel law

The travel law:

$$t(d_{ij}) = \frac{1}{d^2} \qquad \text{(d in minutes)}$$

was assumed the same for all zones.

3. Determining the "generation coefficient" and the traffic on each street.

Population in each zone P_i
Employment in each zone E_j

The traffic between 2 zones was anticipated to be proportional to:

$$T = \frac{P_i \, E_j}{d^2_{ij}}$$

A "trip matrix" was then determined, and this matrix was assigned to the streets using the "all or nothing" principle. For this a speed, based on traffic counts, had to be determined for each street in the network.

The following type of table was then set up:

Street from node	Link to node	1 Counted traffic on street	2 External volume assigned to street	$T = \dfrac{P_i\,E_j}{d^2_{ij}}$ assigned to street	Counted intern. traffic C. 1-2	Estimated internal traffic V.
121	110	15,000	2,300	2,420	12,700	12,860

A total of 23 street links were used. The best formula for V was found by regression analysis to be:

$$V = 730 + 5\,T \quad \text{(correlation coeff. 0.97)}$$

In this equation "5" is approximately what is usually called a generation coefficient.

Low also used population P in both zones in his analysis but in this case the coefficient of correlation was not as high.

II. Abstract of:

Overgaard: Development of a simplified traffic model for the city of Silkeborg. Paper presented at the O.E.C.D. T.7 group in Copenhagen 1972.

1. Determining the external traffic

A roadside interview at 7 streets entering the city determined the external traffic to each of the city's 48 zones. This gave a matrix for the external traffic.

2a. Determining the car density in each zone

The total number of inhabitants, 44,000, and of cars, 11,500 in Silkeborg was known, but not the number of cars in each zone. An investigation in the city of Aarhus had shown that the number of cars per 1,000 inhabitants in single family zones was 50 per cent higher than in zones with apartment houses only. This gave an expected number of 190 cars per 1,000 inhabitants in zones with apartment houses only, and 285 in zones with single family houses only.

2b. Determining the generation of traffic in the city.

From the Aarhus study it was taken that the number of trips per car per day in zones with "single family houses only" is 20 per cent higher than in zones with "apartment houses only". This concerns "home based work trips". For other home based trips this percentage is 50 per cent.

It was anticipated that the average number of trips per car per day was 4.5.

Based on these assumptions the following formula for the traffic T generated (and attracted) to a zone was developed:

$$T = 1.75 \begin{pmatrix} \text{Number of} \\ \text{workplaces} \end{pmatrix} + 0.7 \begin{pmatrix} \text{Number of} \\ \text{inhb.} \end{pmatrix} + 0.008 \begin{pmatrix} \text{Number of} \\ \text{inhb.} \end{pmatrix} X \begin{pmatrix} \% \text{ in} \\ \text{one-family} \\ \text{houses} \end{pmatrix}$$

For 2 zones one with "single family houses only" and one with "apartment houses only" the "traffic calculated" by the formula was compared with the "traffic counted" on all streets entering each of these 2 zones.

The result of this calibration was that the anticipated 4.5 trips per car per day should be raised to 5 trips per car per day as an average for all cars in the city.

3. Determining the travel law

Travel laws with different powers were used to determine the interzonal traffic. The external traffic matrix was added to the internal traffic matrix for the city. The thus calculated total "desire line traffic" was calibrated with the "counted road traffic" across 3 screen lines.

The following travel law was found to be the best:

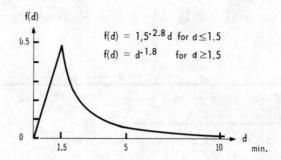

$$f(d) = 1.5^{-2.8} d \quad \text{for } d \le 1.5$$
$$f(d) = d^{-1.8} \quad \text{for } d \ge 1.5$$

The same travel law was used for all zones.

4. Calculation of traffic on each street

The total travel matrices (external traffic included) was then assigned to the streets according to the "all or nothing" method.

III. Abstract of:

T. Jensen and S.K. Nielsen: Calibrating a gravity model and estimating its parameters using traffic volume counts. Proceedings from the English University Traffic Engineer's yearly congress. January 1973.

This model calculates the traffic in a region: The county of Aarhus in Denmark. Size of the area: 5,000 km^2, divided in 25 zones.

1. Determining the external traffic

The rest of the peninsula of Jutland was divided into 9 zones. The traffic from these zones to the county of Aarhus was included in the calculation.

2. Determining the travel law

As a first approximation the simple travel law:

$$t(d) = \frac{1}{d^{\alpha}}$$

was anticipated, with α = 2.0. Later on other values of the power α were tried.

3. Determining the "generation coefficient"

The traffic between 2 zones was anticipated to be proportional to:

$$T = k \frac{P_i \, P_j}{d_{ij}^2} \quad ; \text{ k at first being anticipated to be 1.0}$$

where P = population in the zone.

A "traffic matrix" was thus determined, and the speed on each street based on counted traffic was calculated. This matrix was assigned to the streets using the "all or nothing" principle. The estimated values (V) were then compared with the counted values (C).

Next step was to find the value of k which minimised the expression:

$$Q = \Sigma(V - C)^2$$

It was found that k = 3.4 minimised the expression. The traffic on each street with k = 3.4 was determined.

New speeds were then calculated for each street in the network, corresponding to the calculated traffic volume V.

The traffic matrix was assigned to the network with the new speeds and the value of k, which now minimised the expression above, calculated. The new value of k was found to be 3.6. The new traffic on each street link was determined as "half of the traffic from the final calculation" + "half of the traffic from the new calculation" (Smock's principle: See "An iterative assignment approach", HRR Bulletin 347, 1962). The process was repeated 9 times and the final value of k was 3.8.

4. Continued work on determination of the travel law

The whole procedure above was repeated with other values of the power α : 1.9, 2.1, 2.2, 2.3, and 2.6, and it was shown that α = 2.25 minimised the expression above.

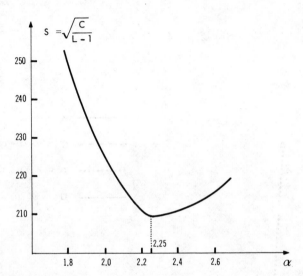

IV. Discussion on the three methods

The methods (Overgaard's excepted) use the same "generation coefficient" for all zones. As it is known that at least in some cities generation coefficients differ quite a lot from zone to zone (in Gävle, Sweden, from 1.4 to 4.2 trips per car per day) there seems to be an advantage in Overgaard's method. Jensen and Nielsen give the following mean errors for the first and ninth step of the iteration:

$$\alpha = 2$$

	k	mean error for calculated traffic
First iteration step	3.4	- 74 cars
Ninth iteration step	3.8	+ 10 cars

This shows that the mean error is much less after the ninth step.

The figure below shows calculation errors for different counted traffic volumes on the roads for each method.

It seems that for traffic volumes from 10,000 to 20,000 cars per day the "Low" and the "Overgaard" methods give about the same errors, 10 per cent to 20 per cent. For traffic volumes below 10,000, the two Danish methods seem to give better results than Low's.

On the figure is also shown the results from a conventional model, based partly on "home interviews" and used for calculation of the traffic on the streets in the city of Aarhus in Denmark. The counted traffic is between 7,000 and 30,000 cars per day. The model in this case gives errors up to 50 per cent on some streets with traffic from 10,000 to 20,000 cars per day. (An assignment procedure with 9 iterations was used in the city of Aarhus.)

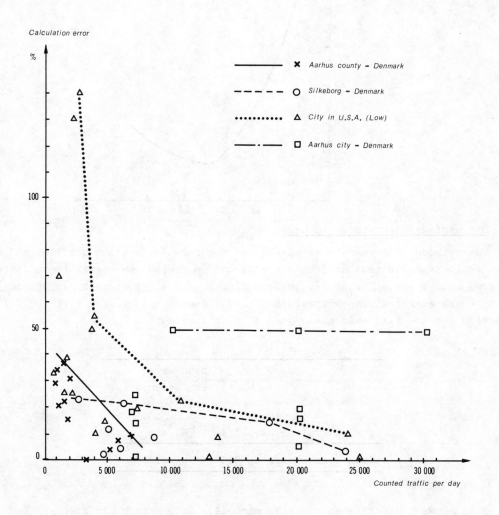

APPENDIX 4

A THEORY OF URBAN ACCESSIBILITY

1.   The idea of accessibility

It is possible to develop the idea of accessibility by empirical reasoning such as
the following:

It is evident that the degree of satisfaction that residents obtain from urban
journeys depends both on the quality of the transport provided and the attractiveness of
the possible destination. This means that with transport of the same quality, attractiveness of the destinations augments according to the variety of choice offered by the
various destinations. For example, the number of jobs vacant reflects the probability
that a resident will find a job that is interesting and well paid; the number of cinemas
reflects the possibility of finding a good film.

A simple way of representing this dual effect (transport quality and variety of
choice) is to count the number of destinations offered for a given need whilst applying
a correcting factor for the distance.

The classic accessibility indicators are thus determined. For a resident in the
zone i looking for a job, for example, the employment accessibility indicator can be expressed as follows:

$$A_i = \sum_j E_j k_{ij}$$

E_j is the number of jobs likely to be available in the zone j.

k_{ij} is a coefficient dependent on the transport time, for example:

$$K_{ij} = e^{-\frac{c_{ij}}{x_o}}$$

The parameter x_o is determined on the basis of traffic survey data and its value
varies according to the person and the trip purpose.

We can see the advantage of the indicator A_i as compared with conventional indicators; not only does it reflect the transport conditions (c_{ij}) but also the variety of
choice offered by the urban structure (E_j).

2.   The theory of urban accessibility

It is possible to found the above empirical reasoning on a theoretical basis. The
economic theory of accessibility developed for this purpose(*) is based on two
assumptions:

(*) See: "The Economic Theory of Urban Accessibility, a Tool for the Urban Planner"
     by G. Koenic, SETRA, June 1974 (an English translation of a shortened version of
     this report can be obtained from O.E.C.D.).

- as assumption regarding the behaviour of the citizen, which optimises the net utility of the transport system. For a citizen looking for a job for example, this net utility resulting from a job vacancy is the difference between the gross utility (annual salary corrected for the interest of the work, social benefits, etc.) and the generalised annual cost of transport (taking account of the time spent);

- an assumption quantifying the probability that the attractiveness of each possible destination is very high, medium or poor(*).

The probable value of the "net utility" can then be calculated for a resident in a given zone using a given means of transport; this net utility is the greatest of the net utilities (for the particular need in question) given the available destinations in the town.

It is significant that the indicator found is precisely the "intuitive" indicator of accessibility; we obtain the following formula for net utility:

$$U_i = x_o \log A_i + \text{constant}$$

where A_i is the accessibility indicator defined above. The service indicator U_i is expressed in economic terms (francs). It takes account of the quality of the transport and the probable desirability of the destinations:

- the transport quality is represented by the journey times or the overall costs of interzone transport c_{ij}, if, for example, all transport costs from a given zone increase by Δc, the net utility for the residents of zone i will be reduced by Δc.

- the probable attractiveness of the destinations is represented by the number of destinations available in each zone (E_j). With the same transport conditions c_{ij}, the greater the number of possible destinations, the more the resident in question is likely to find a destination that appeals to him.

The indicator U_i has two advantages:

(a) Applied to the whole built-up area, it gives an assessment of the overall effectiveness of a given variant. This overall indicator is not subject to the contradictions encountered with conventional indicators such as the total time spent on transport (extending a bus line into the suburbs, by inducing new journeys, can paradoxically increase not only the total time spent on transport but also the satisfaction of the users).

(b) This indicator is also suitable for sectoral analysis, by residential area, mode of transport or by number of cars in use. Thus it is possible to know exactly who benefits or loses by choosing this or that variant; it is also possible to work out the impact of these choices on urban development. Such spatial analysis is especially instructive since large scale transport infrastructures can have substantial - and sometimes unexpected - effects on the whole conurbation.

The theory of accessibility can also be applied directly to traffic forecasting:

(*) This is to consider the gross utility as a random variable for which the probability function must be fixed (the best results are obtained with a negative exponential).

(a) with this theory, it is possible to calculate the probability that a person will find work in a given area; the resulting traffic model is exactly the same as the conventional gravity models, the accessibility theory is thus confirmed on an experimental basis and the gravity models justified theoretically.

(b) it is quite normal that the number of journeys made by residents in a given area depends on the level of service offered; accessibility is thus an explanatory variable in traffic generation. This relationship, unknown in the past due to lack of a suitable indicator for service offered, is confirmed by survey results. Thus an appreciable improvement in traffic generation models may be expected.

There are other possible applications (urban trend models). Nevertheless, it must be noted that this fairly new theory has only just emerged from the research stage.

Comments on figures 1, 2 and 3

The application of the theory of accessibility to the comparison of variants has been illustrated for the case of Le Mans, where a study was carried out to see whether it was preferable to begin by extending the existing Southern ring road to the West, or to build the Northern radial road. For the various alternatives the average value of the net utility indicators have been estimated per zone, in French francs (1970). If a French franc value is given for x_o, then the net utility indicator would be given in French francs. For home to work trips and for the socio-economic class considered here, gross income: 22,500 FF/year, the estimated yearly average value for x_o is 2,200 FF.

Figures 1 and 2 show, for each variant, the maps of net utility offered by jobs accessible by car for households having an average of 1.2 people at work. Although the town centre is the area where traffic conditions are worst, it is also the one offering the greatest utility, because the density of jobs there is very high.

Figure 3 shows the difference between the utilities offered by these two variants and is a clear illustration of the idea of "area of influence" of an infrastructure. It shows that this "area of influence" is quite different from the strip, one or more kilometres at each side, which might have been expected. In fact a new infrastructure can considerably change the traffic flows throughout the built-up area.

For example, the town centre is better served, paradoxically, by the ring road (which does not pass through it) then by the radial road (which does): the ring road alleviates conditions in the town centre by taking a large number of vehicles travelling between peripheral areas which would otherwise have to pass through the town centre. This is the basic reason for the virtually systematic advantage of the ring road over the radial road.

Similarly, it may seem surprising that the only areas where the radial road can compete with the ring road are to the South, although the radial road is to the North: this results from the fact that the ring road drains a great deal of traffic from its Southern extension, thus forming a barrier between the South and the town centre.

Similar maps can be drawn for other modes of transport (public modes of transport, walking, bicycles, etc.) or for different types of citizen (for examples defined in terms of socio-economic classes, income or level of car ownerwhip). In this way it is possible to make a fine analysis of the comparison between different transport systems or even of urbanisation.

Conversely, it is possible to use the theory of accessibility to work out the overall difference in economic efficiency between variants. For workers returning home from work at the evening peak hour, the economic advantage of the ring road variant can be estimated at Frs.8.5 million per year which is considerable.

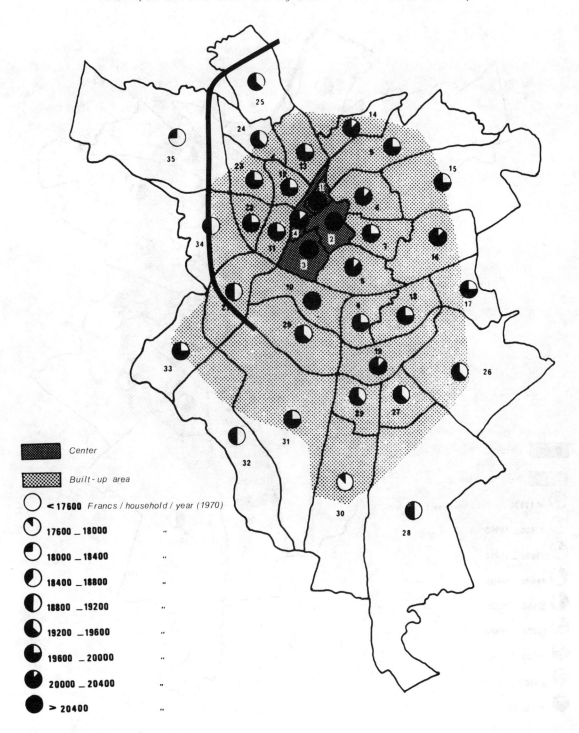

Figure 1

LE MANS – 1977 – PLANNING ALTERNATIVE : RING ROAD

NET UTILITY OFFERED BY JOBS ACCESSIBLE BY CAR

(GROSS INCOME REDUCED BY GENERALISED TRANSPORT COST) FRANCS / HOUSEHOLD / YEAR (1970)

The utility has been calculated for an average household with an income of 22 500 F / year

Center

Built-up area

< 17600 *Francs / household / year (1970)*

17600 _ 18000 ,,

18000 _ 18400 ,,

18400 _ 18800 ,,

18800 _ 19200 ,,

19200 _ 19600 ,,

19600 _ 20000 ,,

20000 _ 20400 ,,

> 20400 ,,

Figure 2

LE MANS – 1977 – PLANNING ALTERNATIVE : RADIAL ROAD

NET UTILITY OFFERED BY JOBS ACCESSIBLE BY CAR
(GROSS INCOME REDUCED BY GENERALISED TRANSPORT COST) FRANCS/HOUSEHOLD/YEAR (1970)

The utility has been calculated for an average household with an income of 22 500 F / year

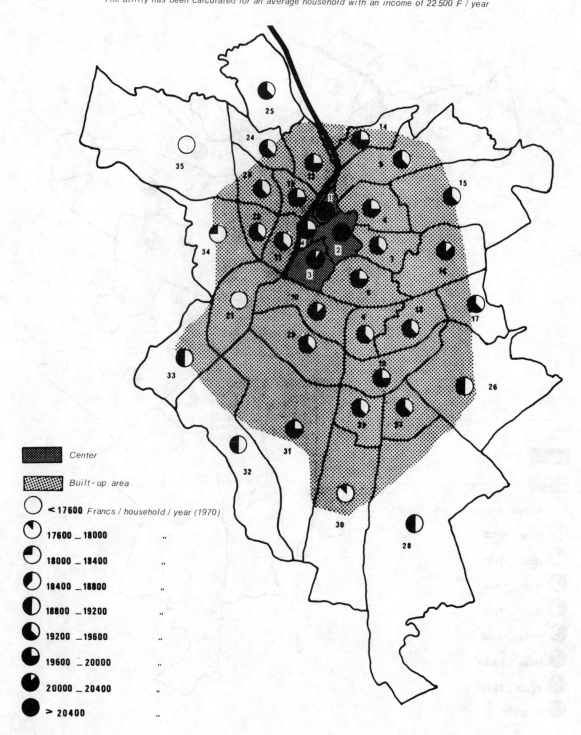

Center

Built-up area

< 17600 *Francs / household / year (1970)*

17600 _ 18000 „

18000 _ 18400 „

18400 _ 18800 „

18800 _ 19200 „

19200 _ 19600 „

19600 _ 20000 „

20000 _ 20400 „

> 20400 „

Figure 3

LE MANS – 1977 – COMPARISON BETWEEN RADIAL ROAD AND RING ROAD

DIFFERENCE IN UTILITY BETWEEN THE TWO SOLUTIONS FOR AN AVERAGE HOUSEHOLD

(In francs / household / year – positive if ring road offers higher utility)

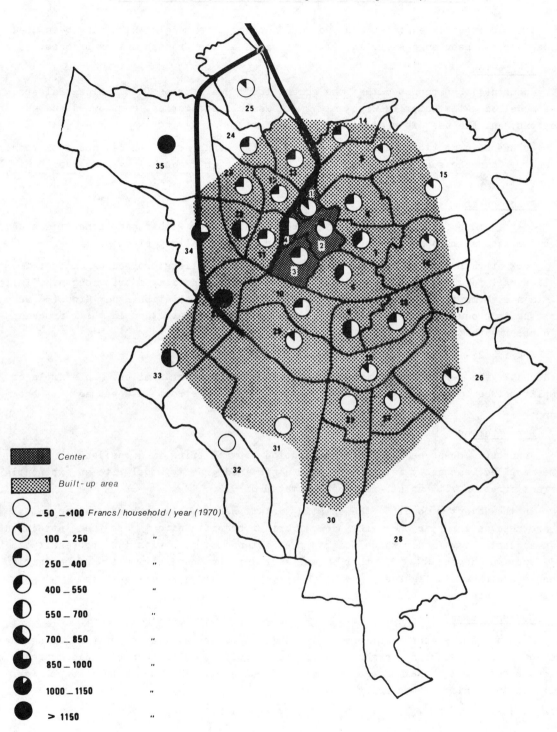

APPENDIX 5

EXAMPLES OF APPROXIMATE METHODS FOR PEDESTRIAN TRIPS

In this appendix examples are quoted of approximate models which have been used, or whose use has been suggested, in different parts of the transportation process.

1.   Generation

Johnson(4) has reported that the generation of walk trips in a residential area may be modelled satisfactorily with only two household categories, cars owned and number of persons, and the use of an accessibility factor.

As has been mentioned earlier, pedestrian trip-ends are often deduced from total trip ends. Thus the specific generation of walk trips is frequently a consequence of modal split.

2.   Modal split

It is common for total trip ends to be known, either within each zone of a study area or as access trips for a single generator such as a railway station.

Work at TRRL has shown that the proportion of travellers proceeding on foot from main-line rail termini may be regarded as an invariable function of trip length. Similarly when, in a much more complex situation, the Coventry Transportation Study Group(3) were attempting to produce a simple walking/cycling model the first stage in their procedure was to observe that the curve

$$\text{Log } Y = 2 - 0.1x^{1.769}$$

reproduced Y, the percentage of travellers still walking after x miles. An example of the split between modes for journeys to the main shopping street in Copenhagen is shown in figure III.

3.   Distribution

Johnson(4) observes that the distribution of walk trips in a residential area can be modelled reasonably well by a doubly contrained gravity model with an exponential deterrent function with an all-purpose parameter of 0.2.

In the Coventry model(3) the existing relationships between homes and potential jobs are used as input to the walk model. Further specific effects are then introduced via an accessibility factor which weights the distance to jobs with the 'propensity to walk' function quoted above. Other examples of data on the distribution of pedestrian distances are shown in figures I and II and results obtained by Mitchell(12) for three English towns are summarised in figure IV.

4.   Concentration

Various simple approximations of the third type are available for predicting the number of people assembling in shopping areas. Percivall and Sandahl(11) discovered that they could represent the simultaneous number of pedestrians in each street length (or matrix) in the centre of Ovebro by:

$$T_i = a.Y_i + b.P_i + c.B_i + d.C_i + e.U_i + f.G_i + g.S_i + h.P_{ki} + k$$

where:

T_i = the simultaneous number of pedestrians in matrix "i".

a = the simultaneous number of pedestrians accumulated around retailing and other commerical service activities during the dimensioning period.

Y_i = effective net floor space area in square metres for retailing or other commercial service activities. Effective net floor area represents total annual turnover expressed as an optimal floor area space.

b = simultaneous number of pedestrians accumulated around parking areas for long term parking.

P_i = number of parking spaces for long term parking.

c = simultaneous number of pedestrians accumulated around bus stops.

B_i = number of bus routes (stopping).

d = total pedestrian accumulation addition on the basis of centrality in town centre.

C_i = proportion of "d" in matrix "i", or centrality of matrix "i".

e = proportion of pedestrian traffic crossing the boundary of the central area expressed as goal orientated movement towards the central area.

U_i = number of pedestrians crossing the central area boundary.

f = simultaneous number of pedestrians accumulated around street stalls.

G_i = number of street stalls or kiosks.

g = simultaneous number of pedestrians accumulated around public seating places.

S_i = number of public seating places.

h = simultaneous number of accumulated pedestrians around public seating spaces.

P_{ki} = number of public parking spaces for short term parking.

k = Constant (intercept). Number of pedestrians in matrix "i" that on the basis of the model chosen, cannot be directly related to any one of the descriptive variables.

They consider simplifications such as

$$T = 0.24 \ Y$$

but find that these lead to a noticeable loss of accuracy.

A simple rule of thumb which belongs either in this section or to section (5) is the observation that pedestrians feel congested when the concentration rises to 0.2 people/m^2(3).

5.   Speed/flow

Many measurements have been made of the average speed/flow or speed/density relationships for pedestrian movement under various conditions (passage ways, stairs, etc.). Values commonly used when a single value is needed are 1.3 m/s for relatively free flow situations and 0.9 m/s when there is moderate congestion.

A simple formula which accounts for the retarding effect of shop frontages on link speed is that used by Ness(7)(8) in a model of central Toronto.

References

1. Barrett, R., "Moving pedestrians in a traffic free environment", Planning and Transportation Research Co. Ltd. Conference 1972, London, 1971.

2. Bruce, J.A., "The pedestrian", Traffic Engineering Handbook, Washington, 1962 (Inst. of Traffic Engineers, pp. 108-141).

3. Edwards, J.A., and Shipley, S., "The Coventry transportation study walk model", Planning and Transportation Research Co. Ltd. Conference, 1972.

4. Johnson, K.R., "A pedestrian model based on a home interview survey", Planning and Transportation Research Co. Ltd. Conference 1972, London, 1971.

5. Ludmann, "Pedestrian Areas in German Cities", Fussgängerbereiche in Deutschen Städten, Köln (Kohlhammer).

6. Navin, F.P.D., and Wheeler, R.J., "Pedestrian flow characteristics", Traffic Engineering, June 1969.

7. Ness, M.P., "Some problems of simulating pedestrian movement", Planning and Transportation Research Co. Ltd., Conference, 1972.

8. Ness, M.P., Morrall, J.F., and Hutchinson, B.G., "An analysis of Central Business District pedestrian circulation patterns", Highway Research Record 283, Washington (1969).

9. O'Flaherty, C.A., and Parkinson, M.H., "Movement in a city centre footway", Traffic Engineering and Control.

10. Older, S.J. "Movement of pedestrians in footways in shopping streets", Traffic Engineering and Control, 1968, 10 (4).

11. Purcivall and Sandahl, J., "Pedestrian traffic forecast model for town centres", Planning and Transportation Research Co. Ltd. Conference 1972, London, 1971.

12. Mitchell, C.G.B., "Pedestrian and cycle journeys in English urban areas", Transport and Road Research Laboratory Report LR 497, 1973.

13. Kanstrup, "The function of the pedestrian area "Strøget" in Copenhagen City", Proceedings Int. Fed. for Housing and Planning Congress, Copenhagen, 1973.

Figure I

DISTRIBUTION OF PEDESTRIAN WALKING DISTANCES

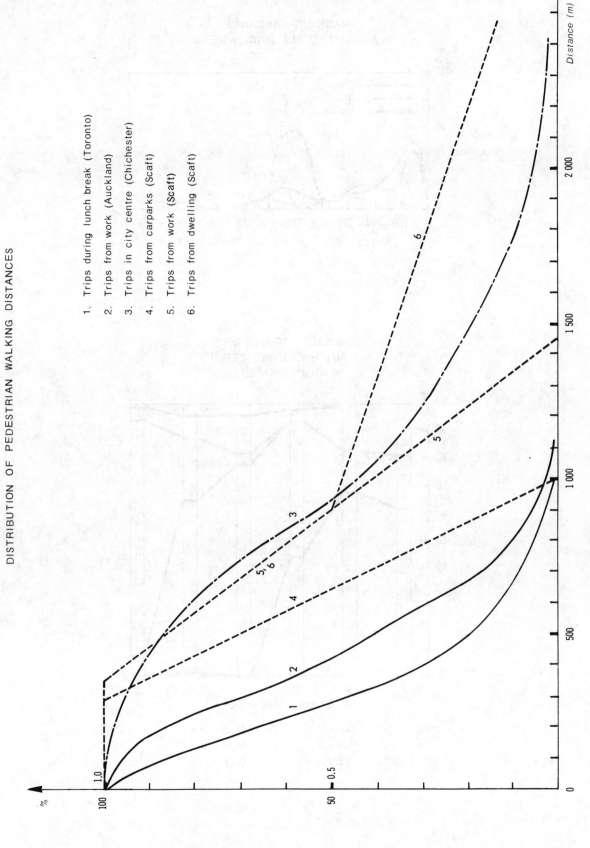

1. Trips during lunch break (Toronto)
2. Trips from work (Auckland)
3. Trips in city centre (Chichester)
4. Trips from carparks (Scaft)
5. Trips from work (Scaft)
6. Trips from dwelling (Scaft)

Figure II

WALKING DISTANCE
ACCEPTED BY PASSENGERS

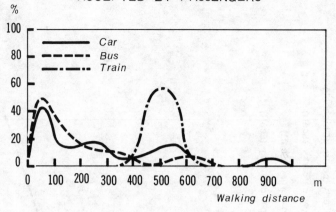

Figure III

MEANS OF TRANSPORT
TO MAIN SHOPPING STREET
IN COPENHAGEN

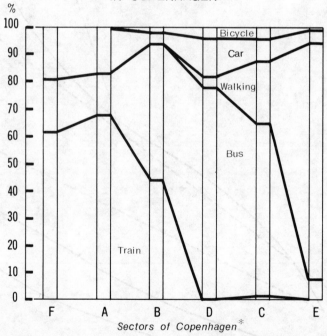

* ORIGIN : F = Other   D = South town
            A = Out town   C = North town
            B = Suburbs   E = City

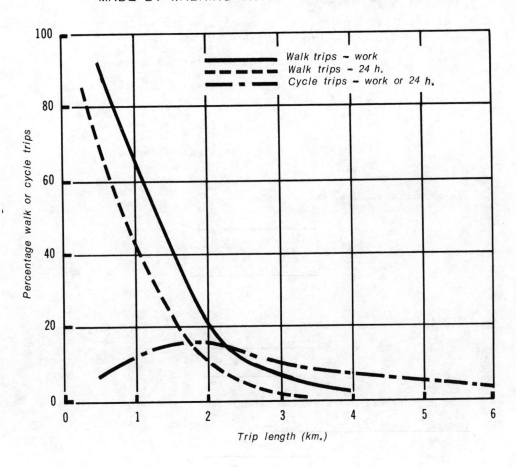

Figure IV

COLLECTED PROPORTIONS OF TOTAL TRIPS
MADE BY WALKING OR ON PEDAL CYCLES

APPENDIX 6

SOME DETAILS ON ERRORS IN MODEL COMPONENTS

1.   Definitions

The degree of error associated with traffic models is expressed by the various authors in different terms. Some of these terms are given below.

A. Difference between observed and calculated mean value, both in absolute terms and in percentage.

B. It is possible to calculate the mean value and to determine how much the highest and lowest values deviate from the mean value.

In the equations below, the following notations are used:

$$Y_s = \text{surveyed value}$$
$$\overline{Y}_s = \text{mean of surveyed values}$$
$$Y_e = \text{estimated value}$$
$$n = \text{number of observations}$$

C. Root Mean Square (RMS) error

The percentage RMS error:

$$= \frac{\sqrt{\dfrac{\sum_n (Y_s - Y_e)^2}{n}}}{\overline{Y}_s} \times 100$$

D. Standard error

$$S = \sqrt{\frac{\sum_n (Y_s - Y_e)^2}{n}}$$

Some authors use n - 1 in the denominator.

E. Standard deviation and variance

$$\sigma = \sqrt{\frac{\sum_n (Y_s - \overline{Y}_s)^2}{n}}$$

Variance $= \sigma^2$

Some authors use n - 1 in the denominator.

F. Coefficient of multiple correlation and coefficient of multiple determination

$$R = \sqrt{1 - \frac{S^2}{\sigma^2}}$$

116

Coefficient of multiple determination = R^2.

R^2 . 100% gives the proportion of the variance, which is explained by the variables.

G. 95 per cent limit for deviation expressed in percentage of the mean value of observations if (say) 1 per cent are surveyed, or as the mean value of all observations if 100 per cent are surveyed.

H. The root mean square error can also be calculated for 100 per cent surveyed and e.g. 1 per cent surveyed. The error in traffic calculations is often related to the traffic volume (e.g., between two zones). With small traffic volumes, the error can be very great. By compounding the small traffic volumes (e.g., calculating the traffic between a greater number of zone), the error becomes smaller.

2. <u>Errors affecting estimates of traffic generated by households - errors due to different sample size</u>

For traffic from an "average houshold" (representing 27,500 <u>London</u> housholds) the figures were as follows(6):

Number of housholds interviewed	55	275	1,650
Per cent of housholds interviewed	0.2	1.0	6.0
Error in traffic generation estimate (95 per cent limit)	±12%	±6%	±2.5%

<u>Sioux Falls</u>(8). 20,000 housholds, not classified.

The following mean errors in the number of trips per household were found according to the indicated home interview sample sizes:

Per cent of households interviewed	1%	12.5%
Errors in traffic generation	±7%	±2%

3. <u>Errors affecting traffic from a zone, due to the inaccuracy of the model</u>

Nordqvist(1) states that if the traffic from a given zone is, in reality, N trips per day, the number of trips calculated with the aid of the model can be expected to range from 0.5 N to 2 N, corresponding to an error percentage of -50 per cent and +100 per cent.

In the case of the Swedish town of Gävle, the generation model is given as 2.8 person trips by car per day. In reality, however, this figure was found to vary between 2.8 ± 1.4 person trips by car per day, corresponding to a variation of ±50 per cent.

<u>Ruske</u>(2): Average for seven German cities (with three types of zones in each city).

By using a model, one obtains the following percentage of the measured trips correctly:

1. for residential areas   94 per cent   (Bestimmtheitsmass)
2. for mixed areas   90 per cent
3. for central areas   62 - 80 per cent

Odense, Denmark, divided into four zones(3).

The theoretically calculated trips were found to differ by ±25 per cent from the trips ascertained from home interviews.

For Seattle it was stated that 96 per cent of the town's traffic in terms of person trips can be explained by floor space relations(4).

For three categories of trips the errors affecting traffic from a given zone due to the inaccuracy of the model:

In the case of 17 French towns:

Home-work	7 per cent
Home-non work	10 per cent
Secondary	15 per cent

For Bangalore(5) the following was obtained:

Home-work	28 per cent RMS
Non home based	40 per cent

4. Errors affecting traffic between two zones

Errors due to sample size. One category of trips

London. Three million households, twelve million trips per day, 1,000 zones(6).

The error varies greatly with the volume of traffic between the two zones. The table below shows the errors (for a 95 per cent confidence limit) in the numbers of trips between two zones for 1,000, 10,000, etc. trips per day. As far as the highest volumes are concerned, these are related not to traffic zones but to traffic districts (i.e. groups of zones):

Trips per day		1,000	10,000	60,000	120,000
Sample size	0.2%	- 90% + 260%	± 45%	± 18%	± 12%
	1.0%	± 60%	± 20%	± 10%	± 6%
	6.0%	± 25%	± 8½%	± 3%	± 3%

Errors due to the inaccuracy of the model. One category of trips

Swiss towns (Zuberbühler)(7)

Trips per day of desire line	50	100	200	400
Maximum difference*	± 170%	± 140%	± 70%	± 20%
Average difference*	± 70%	± 45%	± 22%	± 16%

* Difference between observed and calculated values.

Errors due to the inaccuracy of the model. Three categories of trips

17 French towns

	Error percentage
Home-work	15%
Home-non-work	14%
Secondary	10%

Sioux Falls(8)

	Error percentage	Average traffic
Home-work	18%	500-1500 trips per day
Home-non-work	16%	1000-3000 " " "
Non home based	15%	1000-5000 " " "

Kolding, Denmark(9)

	Error percentage	Average traffic
Home-work	107%	Approx. 100 trips per day
Home-non-work	76%	" " " " "
Non home based	79%	" " " " "

5. Discrepancies between measured and estimated screen line traffic

Errors presumed to be due to the inaccuracy of the model. One category of trips

The patterns of screen line traffic do not differ greatly from that of zone-to-zone traffic, except that the traffic volumes are heavier.

Nordqvist(1) reports errors of 5 to 6 per cent in screen line traffic data obtained in Malmö, Norrköping and San Francisco.

Irish data(10) indicate a screen line discrepancy of 9 per cent in one survey.

For the case of eight screen lines through Odense, Denmark(3), the error was found to be ±5 per cent, with too much traffic on the radial roads and too little on the minor roads.

None of these three surveys state whether the percentage depends on the traffic volume.

At Kolding (Denmark)(9), the screen line data showed the following errors in calculated traffic:

	Error percentage	Traffic volume per day
Section 1	34%	7,600
Section 2	14%	6,000

For Kolding, the mean error is only insignificantly greater when using the three categories separately.

Using a screen line check of car vehicle trips calculated from an 0.25% sample of journeys to work in <u>Copenhagen</u> in 1956, the following error percentages were found(11):

Screen line No.	Traffic calculated from model. Percentage deviation	Number of cars per day
1	0	40,000
2	3	31,800
3	0	130,000
4	-40	44,000
5	2	96,000
6	+47	43,000
7	+30	14,000
8	+25	10,000

<u>Check of screen line traffic. Several categories of trips</u>

<u>Sioux Falls</u>(8)

Screen line No.	Person trips per day	Calculated from the model after conversion with the aid of empirically found "constant numbers of trips by car per day" and an empiric "travel function." Percentage deviation.		
		1	3	6 categories
1	2,300	+22%	18%	+17%
2	7,900	-12%	- 8%	- 6%
3	21,000	- 2%	- 2%	- 2%

6. <u>Errors affecting forecasts</u>

<u>Investigation of the reliability of the growth factor method, once the real growth factors are known</u>

In "Public Roads," 1958(12), a survey is described where the 1948 traffic in Washington D.C., is updated to 1955 by using the growth factors actually experienced in the different zones, and is then compared with the actual 1955 traffic.

In the time from 1948 to 1955, the following percentage increases were experienced:

Number of cars	96%
Number of passenger car trips	89%
Number of motor vehicle trips	67%
Number of taxi trips, no more than	8%

For <u>zone-to-zone</u> passenger car trips, using 254 zones, the root mean square error was on average 138 per cent. There are many zone-to-zone pairs with very few trips, and the errors affecting the small numbers of trips are very great. If the number of zones is reduced to 67, the average error for zone-to-zone trips is reduced to 35 per cent.

120

The RMS error depends greatly on the traffic volume between the two zones.

Daily traffic	RMS error
100 cars	70%
1,000 "	30%
10,000 "	15%
50,000 "	6%
120,000 "	5%

It is not clear whether these data apply to the desire line traffic between two zones or to traffic on the road after assignment.

Investigation of errors in calculated road traffic due to errors in forecasting the population and car ownership trends

Strand(13) assumes that the growth factor method yields a correct distribution, and that the assignment has been carried out correctly. The error affecting traffic counts in 1965 is estimated at 15 per cent.

If the population growth from 1965 to 1975 is estimated at 10 ± 50 per cent, i.e., at 5 to 15 per cent, the error affecting the 1975 road traffic forecast is calculated to be 15.7 per cent.

If the growth in the car ownership rate from 1965 to 1975 is estimated at 150 ± 50 per cent, i.e., at 75 to 225 per cent, the error affecting the 1975 road traffic forecast is calculated to be 33.5 per cent. A very high growth in car ownership has thus been assumed.

If the car ownership forecasts are combined with a 1965-1975 population growth forecast ranging from 5 to 15 per cent, the error affecting the 1975 road traffic forecast becomes 33.8 per cent. Errors in respect of population growth play a very small part.

In this investigation, the error percentage was not related to the traffic volume.

References

1. Nordqvist, "Studies in traffic generetics", National Swedish Building Research, Doc. No. 2, 1969.

2. Ruske, "Stochastische und deterministische Modelle zur Errechnung des Verkehrsaufkommens aus Strukturmerkmalen", Dissertation, Aachen, 1968.

3. Danish Road Directorate, "Trafikanalyse i Odense - Datarapport", 1969.

4. Harper and Edwards, "Generation of persons". Highway Research Record, No. 253.

5. Srinivanzan, "Study on urban travel characteristics". Central Road Research Institute, Delhi. Road Research Paper No. 101, 1969.

6. Crawford, "Accuracy of the transportation model". Greater London Council Transportation Research, Research Memorandum, 1968.

7. Zuberbühler, "Technische Aspekte Städtischer Verkehrsplanungen", Automobilismo e Automobilismo Industriale, No. 5-6/1971.

8. Ben, Bouchard and Sweet "An evaluation of simplified procedures for determing travel patterns in a small urban area". Highway Research Record No. 88, 1965.

9. Danish Road Directorate, "Færdselsundersøgelse i Kolding - Datarapport", 1969.

10. Dublin Transportation Study: Technical Report No. 16, paragraph 16.25, An Foras Forbatha, Dublin, 1973.

11. Svejstrum and N.O. Jørgensen, "Trafikprognoser i byer", Stads- og Havneingeniøren, No. 4, Copenhagen, 1968.

12. Brokke and Mertz, "Evaluating trip forecast methods", Public Roads, October 1958.

13. Strand, "The stepwise planning procedure". Traffic Quarterly, January 1972.

MEMBERS OF THE GROUP

Chairman: Mr. A.R. Cawthorne, United Kingdom

Technical Secretaries: Mr. J. Devlin, Ireland
                                 Mr. G. Koenig, France.

AUSTRIA   Mr. H. Berger
                                 Federal Ministry for Trade, Commerce and Industry
                                 Wien

BELGIUM   Mr. C. Couvreur and Mr. Tilmans
                                 Service du Trafic Routier
                                 550 Chaussée de Louvain
                                 1030 Bruxelles

                                 Mr. G. Hoste
                                 Service Promotion des Transports
                                 Urbains du Ministère des Communications
                                 2, Place St. Lazare
                                 1030 Bruxelles

                                 Mr. C. Rochez
                                 SOBEMAP
                                 Place du Champ de Mars, 5,
                                 B 1050 Bruxelles.

CANADA   Mr. B.R. McKeown
                                 Planning Branch
                                 Department of Highways
                                 Victoria
                                 British Columbia

DENMARK   Professor P.H. Bendtsen
                                 The Technical University of Denmark
                                 Department for Road Construction
                                 Transportation Engineering and Town Planning
                                 DTH Bygning 115
                                 2800 Lyngby

                                 Mr. Søren Budde
                                 Institut for Vejbygning
                                 DTH, Bygning 115
                                 2800 Lyngby

FINLAND   Mr. H. Turunen
                                 Liikennetekniikka
                                 Itälahdentie 20
                                 00210 Helsinki 21

FRANCE   Mr. G. Koenig
                                 S.E.T.R.A.
                                 46 Avenue A. Briand
                                 92 - Bagneux

FRANCE(cont'd)	Mr. Seigner S.E.T.R.A. 46 Avenue A. Briand 92 - Bagneux
GERMANY	Mr. Busch Ministerialrat Bundesverkehrsministerium Abteilung Strassenbau 53 Bonn Sternstrasse 100
	Dr. Ruske Technische Hochschule Institut für Stadtbauwesen Mies-van-der-Rohe-Str. 51 Aachen
	Mr. P. Züll Bundesverkehrsministerium 53 Bonn Sternstrasse 100
IRELAND	Mr. J. Devlin An Foras Forbartha St. Martin's House Waterloo Road Dublin 4
JAPAN	Mr. Kazuo Yoda Deputy Head City Planning Division City Bureau Ministry of Transport 2-1-3 Kasumigaseki Chiyoda-Ku Tokyo
	Mr. Choji Tomita Senior Policy Planning Officer Secretariat to the Minister Ministry of Transport 2-1-3 Kasumigaseki Chiyoda-Ku Tokyo
SPAIN	D. Francisco Mir-Espinet Dr. Ingeniero Industrial Delegacion de Circulacion y Transportes Ayuntamiento de Barcelone
	D. Antonio Valdés y Gonzáles-Roldán Dr. Ingeniero de Caminos Ministerio de Obras Publicas Avda del Generalisimo 3 Madrid 3

SWEDEN	Mr. K. Sicking Statens Vägverk, Fack 10220 Stockholm 12
SWITZERLAND	Mr. C. Zuberbühler c/o Seiler Niederhauser Ing. Büro AG Jungholzstrasse 6 CH-8050 Zürich
UNITED KINGDOM	Mr. A.R. Cawthorne Transport and Road Research Laboratory Old Wokingham Road Crowthorne Berkshire
UNITED STATES	Dr. B. Dial Office of Research Development and Demonstration Urban Mass Transportation Administration Washington DC 20008
O.E.C.D.	Professor M.L. Manheim Consultant to the O.E.C.D. Environment Directorate Transportation Systems Division Dept. of Civil Engineering Room 1-138, M.I.T., Cambridge Mass. 02139
	Mr. B. Horn O.E.C.D. Road Research Secretariat

The Editing Committee was composed of:

Messrs.   Cawthorne
Devlin
Horn
Koenig
Seigner

ABSTRACT

Urban traffic is one of the crucial problems of governmental transport policies. In order to make urban transport investments and policies more effective, it is necessary to analyse and evaluate a wide range of alternative transport solutions before selecting those most adequate for detailed study. The aim of this Report is therefore to appraise and improve existing decision making tools available in the form of "strategic" transport computer models and to study the degree of simplification which is compatible with meaningful results.

The Report reviews the current state of the art in urban traffic models, i.e. the four-step models (generation, distribution, modal split, assignment) conventionally used in urban transportation and the new single-step "explicit" demand/supply equilibrium models; the concept of the General Share Model is introduced and the various modelling approaches are discussed. The implications for simplified traffic models and current research in this field are described. The Report also contains an evaluation of the main issues involved in modern urban transport planning and of the major model components such as changes in land use, public transport, new modes, traffic restraint, parking and pedestrians. A short chapter on accuracy considerations is included and presents some practical results of past transportation studies. Finally data requirements, availability and simplification are examined.

The report concludes with a series of recommendations for transport planners and proposals for further research in this area.

LIST OF PUBLICATIONS OF THE ROAD RESEARCH PROGRAMME

Road Safety

    Alcohol and drugs (January 1968)
    Pedestrian safety (October 1969)
    Driver behaviour (June 1970)
    Proceedings of the symposium on the use of statistical methods in the analysis
    of road accidents (September 1970)
    Lighting, visibility and accidents (March 1971)
    Research into road safety at junctions in urban areas (October 1971)
    Road safety campaigns: design and evaluation (December 1971)
    Speed limits outside built-up areas (August 1972)
    Research on traffic law enforcement (April 1974)

Road Traffic

    Electronic aids for freeway operation (April 1971)
    Area traffic control systems (February 1972)
    Optimisation of bus operation in urban areas (May 1972)
    Two lane rural roads: road design and traffic flow (July 1972)
    Traffic operation at sites of temporary obstruction (February 1973)
    Effects of traffic and roads on the environment in urban areas (July 1973)
    Proceedings of the symposium on techniques of improving urban conditions by
    restraint of road traffic (September 1973)

Road Construction

    Research on crash barriers (February 1969)
    Motor vehicle corrosion and influence of de-icing chemicals (October 1969)
    Winter damage to road pavements (May 1972)
    Accelerated methods of life-testing pavements (May 1972)
    Proceedings of the symposium on the quality control of road works (July 1972)
    Waterproofing of concrete bridge decks (July 1972)
    Optimisation of road alignment by the use of computers (July 1973)
    Water in roads: prediction of moisture content in road subgrades (August 1973)
    Maintenance of rural roads (August 1973)
    Water in roads: methods for determining soil moisture content and pore water
    tension (December 1973)

OECD SALES AGENTS
DEPOSITAIRES DES PUBLICATIONS DE L'OCDE

ARGENTINA – ARGENTINE
Carlos Hirsch S.R.L.,
Florida 165, BUENOS-AIRES.
☎ 33-1787-2391 Y 30-7122

AUSTRALIA – AUSTRALIE
B.C.N. Agencies Pty, Ltd.,
161 Sturt St., South MELBOURNE, Vic. 3205.
☎ 69.7601
658 Pittwater Road, BROOKVALE NSW 2100.
☎ 938 2267

AUSTRIA – AUTRICHE
Gerold and Co., Graben 31, WIEN 1.
☎ 52.22.35

BELGIUM – BELGIQUE
Librairie des Sciences
Coudenberg 76-78, B 1000 BRUXELLES 1.
☎ 13.37.36/12.05.60

BRAZIL – BRESIL
Mestre Jou S.A., Rua Guaipá 518,
Caixa Postal 24090, 05089 SAO PAULO 10.
☎ 256-2746/262-1609
Rua Senador Dantas 19 s/205-6, RIO DE
JANEIRO GB. ☎ 232-07. 32

CANADA
Information Canada
171 Slater, OTTAWA. KIA 0S9.
☎ (613) 992-9738

DENMARK – DANEMARK
Munksgaards Boghandel
Nørregade 6, 1165 KØBENHAVN K.
☎ (01) 12 69 70

FINLAND – FINLANDE
Akateeminen Kirjakauppa
Keskuskatu 1, 00100 HELSINKI 10. ☎ 625.901

FRANCE
Bureau des Publications de l'OCDE
2 rue André-Pascal, 75775 PARIS CEDEX 16.
☎ 524.81.67
Principaux correspondants :
13602 AIX-EN-PROVENCE : Librairie de
l'Université. ☎ 26.18.08
38000 GRENOBLE : B. Arthaud. ☎ 87.25.11
31000 TOULOUSE : Privat. ☎ 21.09.26

GERMANY – ALLEMAGNE
Verlag Weltarchiv G:m.b.H
D 2000 HAMBURG 36, Neuer Jungfernstieg 21
☎ 040-35-62-501

GREECE – GRECE
Librairie Kauffmann, 28 rue du Stade,
ATHENES 132. ☎ 322.21.60

ICELAND – ISLANDE
Snaebjörn Jónsson and Co., h.f.,
Hafnarstræti 4 and 9, P.O.B. 1131,
REYKJAVIK. ☎ 13133/14281/11936

INDIA – INDE
Oxford Book and Stationery Co.:
NEW DELHI, Scindia House. ☎ 47388
CALCUTTA, 17 Park Street. ☎ 24083

IRELAND – IRLANDE
Eason and Son, 40 Lower O'Connell Street,
P.O.B. 42, DUBLIN 1. ☎ 01-41161

ISRAEL
Emanuel Brown :
35 Allenby Road, TEL AVIV. ☎ 51049/54082
also at :
9, Shlomzion Hamalka Street, JERUSALEM.
☎ 234807
48 Nahlath Benjamin Street, TEL AVIV.
☎ 53276

ITALY – ITALIE
Libreria Commissionaria Sansoni :
Via Lamarmora 45, 50121 FIRENZE. ☎ 579751
Via Bartolini 29, 20155 MILANO. ☎ 365083
Sous-dépositaires:
Editrice e Libreria Herder,
Piazza Montecitorio 120, 00186 ROMA.
☎ 674628
Libreria Hoepli, Via Hoepli 5, 20121 MILANO.
☎ 865446
Libreria Lattes, Via Garibaldi 3, 10122 TORINO.
☎ 519274
La diffusione delle edizioni OCDE è inoltre assicu-
rata dalle migliori librerie nelle città più importanti.

JAPAN – JAPON
OECD Publications Centre,
Akasaka Park Building,
2-3-4 Akasaka,
Minato-ku
TOKYO 107. ☎ 586-2016
Maruzen Company Ltd.,
6 Tori-Nichome Nihonbashi, TOKYO 103,
P.O.B. 5050, Tokyo International 100-31.
☎ 272-7211

LEBANON – LIBAN
Documenta Scientifica/Redico
Edison Building, Bliss Street,
P.O.Box 5641, BEIRUT. ☎ 354429 – 344425

THE NETHERLANDS – PAYS-BAS
W.P. Van Stockum
Buitenhof 36, DEN HAAG. ☎ 070-65.68.08

NEW ZEALAND – NOUVELLE-ZELANDE
The Publications Officer
Government Printing Office
Mulgrave Street (Private Bag)
WELLINGTON, ☎ 46.807
and Government Bookshops at
AUCKLAND (P.O.B. 5344). ☎ 32.919
CHRISTCHURCH (P.O.B. 1721). ☎ 50.331
HAMILTON (P.O.B. 857). ☎ 80.103
DUNEDIN (P.O.B. 1104). ☎ 78.294

NORWAY – NORVEGE
Johan Grundt Tanums Bokhandel,
Karl Johansgate 41/43, OSLO 1. ☎ 02-332980

PAKISTAN
Mirza Book Agency, 65 Shahrah Quaid-E-Azam,
LAHORE 3. ☎ 66839

PHILIPPINES
R.M. Garcia Publishing House,
903 Quezon Blvd. Ext., QUEZON CITY,
P.O. Box 1860 – MANILA. ☎ 99.98.47

PORTUGAL
Livraria Portugal,
Rua do Carmo 70-74. LISBOA 2. ☎ 360582/3

SPAIN – ESPAGNE
Libreria Mundi Prensa
Castelló 37, MADRID-1. ☎ 275.46.55
Libreria Bastinos
Pelayo, 52, BARCELONA 1. ☎ 222.06.00

SWEDEN – SUEDE
Fritzes Kungl. Hovbokhandel,
Fredsgatan 2, 11152 STOCKHOLM 16.
☎ 08/23 89 00

SWITZERLAND – SUISSE
Librairie Payot, 6 rue Grenus, 1211 GENEVE 11.
☎ 022-31.89.50

TAIWAN
Books and Scientific Supplies Services, Ltd.
P.O.B. 83, TAIPEI.

TURKEY – TURQUIE
Librairie Hachette,
469 Istiklal Caddesi,
Beyoglu, ISTANBUL, ☎ 44.94.70
et 14 E Ziya Gökalp Caddesi
ANKARA. ☎ 12.10.80

UNITED KINGDOM – ROYAUME-UNI
H.M. Stationery Office, P.O.B. 569, LONDON
SE1 9 NH, ☎ 01-928-6977, Ext. 410
or
49 High Holborn
LONDON WC1V 6HB (personal callers)
Branches at: EDINBURGH, BIRMINGHAM,
BRISTOL, MANCHESTER, CARDIFF,
BELFAST.

UNITED STATES OF AMERICA
OECD Publications Center, Suite 1207,
1750 Pennsylvania Ave, N.W.
WASHINGTON, D.C. 20006. ☎ (202)298-8755

VENEZUELA
Libreria del Este, Avda. F. Miranda 52,
Edificio Galipán, Aptdo. 60 337, CARACAS 106.
☎ 32 23 01/33 26 04/33 24 73

YUGOSLAVIA – YOUGOSLAVIE
Jugoslovenska Knjiga, Terazije 27, P.O.B. 36,
BEOGRAD. ☎ 621-992

Les commandes provenant de pays où l'OCDE n'a pas encore désigné de dépositaire
peuvent être adressées à :
OCDE, Bureau des Publications, 2 rue André-Pascal, 75775 Paris CEDEX 16
Orders and inquiries from countries where sales agents have not yet been appointed may be sent to
OECD, Publications Office, 2 rue André-Pascal, 75775 Paris CEDEX 16

OECD PUBLICATIONS, 2, rue André-Pascal, 75775 Paris Cedex 16 - No. 33545 - 1974
PRINTED IN FRANCE